STP 1161

Leak Detection for Underground Storage Tanks

Philip B. Durgin and Thomas M. Young, editors

ASTM Publication Code Number (PCN)
04-011610-65

ASTM
1916 Race Street
Philadelphia, PA 19103

Library of Congress Cataloging-in-Publication Data

Leak detection for underground storage tanks / Philip B. Durgin and Thomas M. Young, editors.
(STP : 1161)
"ASTM publication code number (PCN) 04-011610-65"
Papers presented at the symposium of the same name, held in New Orleans, LA on 29 Jan. 1992.
Includes bibliographical references and indexes.
ISBN 0-8031-1858-9
1. Petroleum products--Underground storage--Congresses. 2. Leak detectors--Congresses. I. Durgin, Philip B. II. Young, Thomas M., 1962-   . III. Series: ASTM special technical publication ; 1161.
TP692.5.L397   1993
665.5'42'0287--dc20   93-14690
                                                                                 CIP

Copyright © 1993 AMERICAN SOCIETY FOR TESTING AND MATERIALS, Philadelphia, PA. All rights reserved. This material may not be reproduced or copied, in whole or in part, in any printed, mechanical, electronic, film, or other distribution and storage media, without the written consent of the publisher.

Photocopy Rights

Authorization to photocopy items for internal or personal use, or the internal or personal use of specific clients, is granted by the AMERICAN SOCIETY FOR TESTING AND MATERIALS for users registered with the Copyright Clearance Center (CCC) Transactional Reporting Service, provided that the base fee of $2.50 per copy, plus $0.50 per page is paid directly to CCC, 27 Congress St., Salem, MA 01970; (508) 744-3350. For those organizations that have been granted a photocopy license by CCC, a separate system of payment has been arranged. The fee code for users of the Transactional Reporting Service is 0-8031-1858-9/93 $2.50 + .50.

Peer Review Policy

Each paper published in this volume was evaluated by three peer reviewers. The authors addressed all of the reviewers' comments to the satisfaction of both the technical editor(s) and the ASTM Committee on Publications.

The quality of the papers in this publication reflects not only the obvious efforts of the authors and the technical editor(s), but also the work of these peer reviewers. The ASTM Committee on Publications acknowledges with appreciation their dedication and contribution to time and effort on behalf of ASTM.

Printed in Philadelphia, PA
March 1993

Foreword

This publication, *Leak Detection for Underground Storage Tanks*, contains papers presented at the symposium of the same name, held in New Orleans, LA on 29 Jan. 1992. The symposium was sponsored by ASTM Committee E-50 on Environmental Assessment. Philip B. Durgin of Veeder-Root in Simsbury, CT and Thomas Young of Ann Arbor, MI presided as symposium co-chairmen and are editors of the resulting publication.

Contents

Overview—P. B. DURGIN … vii

INTERNAL MONITORING

Volumetric Leak Detection—A Systems Perspective—W. F. ROGERS … 3

Error Sources in Automatic Tank Gauging Systems—D. W. FLEISCHER … 17

Leak Detection Methods for Airport Hydrant Systems—J. D. FLORA, JR., W. D. GLAUZ, AND G. J. HENNON … 30

Location of Leaks in Pressurized Petroleum Pipelines by Means of Passive-Acoustic Sensing Methods—E. G. ECKERT, J. W. MARESCA, JR., R. W. HILLGER, AND J. J. YEZZI … 53

EXTERNAL MONITORING

Analysis of UST Leak Vapor Diffusion and Liquid Build-Up—R. P. SCHREIBER AND M. S. ROSENBERG … 73

Evaluation of Metal Oxide Semiconductor and Polymer Adsorption Gas Sensors as Applied to Underground Storage Tank Leak Detection—M. A. PORTNOFF, R. GRACE, A. M. GUZMAN, AND J. HIBNER … 90

Fiber Optic Chemical Sensors–An Overview—A. E. GREY AND J. K. PARTIN … 105

Field Results of Hydrocarbon Vapor Monitoring to Detect Leaking Tanks—P. B. DURGIN AND R. W. MICHELSON … 115

New Vapor Method Detects and Locates Leaks from Pipelines—M. V. MARTIN … 123

Pipeline Leak Detection Using Volatile Tracers—G. M. THOMPSON AND R. D. GOLDING … 131

Regulations and Standards

How Well Do Leak Detection Methods Work?: Preliminary Results from the EPA Test Procedures—T. M. YOUNG — 139

Evaluation of Pipeline Leak Detection Systems—W. D. GLAUZ, J. D. FLORA, AND G. J. HENNON — 151

Expedited Enforcement of UST Regulations in New Mexico—S. A. SUTTON-MENDOZA — 162

Impact of Standards and Certification on Environmental Impairment Liability Insurance Programs—W. P. GULLEDGE — 167

Site and Risk Evaluation

Characteristics of Non-Petroleum Underground Storage Tanks—R. W. HILLGER, J. W. STARR, M. P. MacARTHUR, AND J. W. MARESCA, JR. — 175

Risk Assessment to an Integrated Planning Model for UST Programs—K. W. FERGUSON — 189

Use of On-Site Vapor Analysis in UST Site Assessments: A Summary of Results at 635 Sites—R. D. GOLDING AND T. A. WICHMAN — 197

Screening Methodology for Selecting Clean-Up Technologies at Leaking Underground Storage Tank Sites—C.-Y. FAN AND A. N. TAFURI — 211

Author Index — 229

Subject Index — 231

Overview

The environmental decade of the 1980's brought with it a steady growth in the number and scope of environmental regulations. Much of the concern was directed at the contamination of groundwater supplies by organic chemicals. A newly-emerged, widespread concern was protection of groundwater supplies from underground storage tanks (UST) that leaked fuel. The public realized that the problem might be as close as their corner gas station or even the heating-oil tank buried in their backyard. The potential carcinogenic effects of gasoline components (particularly benzene), that partially dissolve in ground water, heightened the public's anxiety. As a result, there were demands for owners and operators of underground storage tanks to conduct leak tests, provide assurances that their subsurface tanks and pipelines were tight, and clean up sites that had become contaminated by fuels. EPA drafted regulations in response to these demands and they became effective in December 1988.

A considerable amount of research, discussion, and decision-making was devoted to leak detection issues in developing the federal UST regulations. These, together with newly-developed state regulations, have continued to generate strong interest in the private sector from owner/operators who are being regulated as well as from vendors of leak detection equipment and services. Representatives from these and other constituencies joined together to participate in an ASTM subcommittee dealing with leak detection for underground storage tanks. They formed task groups that dealt with leak detection methods outside as well as inside an underground storage tank. Together, they developed an ASTM guide and practice dealing with these issues.

The subject of UST leak detection is interdisciplinary and, as such, has attracted specialists from a variety of disciplines. These include: environmental engineering, chemistry, electronics, groundwater geology, mechanical/electrical engineering, regulatory management, etc. Many of these workers have completed research and reported on it at conferences or in publications sponsored by their own particular specialty. However, until this ASTM conference there had been no conference that focused simply on UST leak detection.

A primary goal of the ASTM Symposium on Leak Detection for Underground Storage Tanks, held in New Orleans in January of 1992, was to bring together UST leak-detection specialists for participation in a forum that would generate a publication where readers could have important UST research under one cover. The papers provide a state-of-the-art review to many leak detection issues. In some cases, the papers report on research that was conducted two or three years ago but has never been adequately directed to the UST leak-detection audience. In other cases, the papers report on the latest UST research. Much of the leak detection research has been sponsored by the USEPA.

Although the amount of leak-detection research, conducted by industry and government, has expanded dramatically since the release of EPA's regulations, conclusive answers to some research questions remain elusive. The papers in this volume represent the most up-to-date review of this research and have been peer reviewed to insure that unsupported statements conflicting with the consensus of opinion among leak-detection experts were omitted. Nevertheless, legitimate differences of opinion about leak-detection methods persist in those areas lacking definitive research results. The editors have chosen to include such conflicting opinions so that readers may reach their own informed conclusions on these issues.

The phased-in approach of the UST regulations guarantee that owners and operators of UST systems will continue to need answers on how to deal with the regulations at least until 1998. Many simply will want to know the regulatory requirements and leak detection equipment on the market. However, for others this volume is intended to provide an objective, in-depth view of several UST issues. Regulators and vendors should also have an interest in this volume.

EPA developed the UST regulations with an eye towards allowing and promoting future improvements in leak-detection equipment and procedures. Conferences and volumes such as this help to communicate the issues and act as a catalyst for further development of UST leak-detection technology. The Symposium was divided into four sessions that were entitled: I. Internal Monitoring, II. External Monitoring, III. Risk, Reliability, & Regulation, and IV. Site Evaluation. There was also a keynote speech by David Ziegele, Director of EPA's Office of Underground Storage Tanks. The title of his speech was "Speeding Up UST Site Assessment and Remediation: EPA's View". He addressed the increasing number of confirmed UST releases (170,000) and how EPA is trying to streamline the regulatory process and get cleanups started early with new, innovative approaches. While his paper is not in this volume a few additional papers, not presented at the Symposium, are published here.

Internal Monitoring

The book opens with a discussion of statistical inventory reconciliation (SIR) by Warren Rogers. This approach to UST leak detection deserves close attention because it is a relatively low-cost alternative with no equipment needs. SIR is receiving increased attention with the number of supporters and detractors both rising. Rogers makes the basic and important point that determination of a leak rate is related to the time between observations of fluid at rest as well as the precision of an observation. In other words, the shorter time that you look at a tank level the more precise you need to be for leak detection. The next paper, by Don Fleischer, is an enlightening look at the errors associated with internal monitoring of tanks when using automatic tank gauges. This information, based on several years of UST testing, demonstrates how such errors can occur and puts them in proper perspective.

Flora, Glauz, and Hennon provide an excellent, comprehensive overview (including estimated costs) of the various leak-detection options for airport hydrant systems. Clearly the greatest problem at airports is dealing with relatively large, long, pressurized pipelines rather than the tanks per se. Maresca and Eckert examine one approach to this problem in their research paper that applies acoustic sensing to the location of leaks in pressurized pipelines.

External Monitoring

UST leaks can also be detected by monitoring the environment surrounding the tanks. The two primary methods are to monitor for hydrocarbon vapors in the soil gas or to detect fuel floating on top of

OVERVIEW   ix

the ground water. External monitoring is dependent upon the capability of the sensor, their number, location, and analysis of the data they generate. External monitoring is greatly complicated by the fact that each site has its own specific characteristics.

Schreiber and Rosenberg report how soil gas and groundwater conditions, following injection of hydrocarbons, can be modelled with computer programs. They demonstrate that diffusion of hydrocarbon vapors in a sandy backfill is influenced by temperature, soil moisture, and the type of organic compound. Their results have been validated with a large, outdoor physical model by Richard Johnson of the Oregon Graduate Institute. Schreiber and Rosenberg also demonstrate that computer modelling can address how fuel drains down to the water table and spreads out upon it.

The hardware for external monitoring is obviously important. Portnoff et al. present the results of research comparing the two types of vapor sensors that are commercially available for monitoring at UST sites. Each type has its own advantages and disadvantages. The review paper by Grey provides information on fiber optic sensors that have the future potential of providing more accurate data for monitoring hydrocarbon vapors as well as fuel dissolved in ground water.

The paper by Durgin and Michelson provides some field results using various types of vapor sensors. Published continuous vapor data and analysis from field sites has been sorely lacking. Their conclusion is that variations in vapor concentrations are real and explainable but there are so many that automatic data analysis becomes necessary.

External monitoring has also addressed the issue of pipeline leakage. Martin and Jensen provide information on how leaks can be detected and their location determined by pulling a vacuum in a permeable tube laid along the pipeline. Thompson and Golding present a similar approach but describe how tracer chemicals can provide additional information.

Regulations and Standards

This section of the book deals primarily with protocols, regulations, and standards that help maintain quality in UST leak detection. Young provides an overview of how the USEPA set up third-party testing of leak-detection equipment/methods and the reasoning behind it. Glauz et al. focuses on EPA's test protocol for pipeline leak detection and suggests revisions for improvement.

While UST leak detection is required throughout the U.S. there are questions about the level of compliance. Sutton-Mendoza demonstrates how New Mexico has taken the UST regulations to the field, enforced them, and quantified their success in expediting enforcement. There is a need to have similar types of information from other states as the USEPA proceeds with encouraging enforcement of leak detection throughout the country. Gulledge addresses how performance criteria and standards, such as those from ASTM, can influence insurance programs in both a positive and negative manner.

Site and Risk Evaluation

One of the driving forces behind leak-detection monitoring is the desire to reduce the financial risk incurred when tank leakage contaminates the subsurface. There are a variety of factors that can increase the risk at a site. One of these is the type of chemical in the tank and Hillger et al. provide an informative survey of the chemicals that are stored in UST.

Several organizations have faced the task of having to deal with a large number of USTs yet not knowing where to start. Ferguson demonstrates how the U.S. Postal Service assessed the environmental and regulatory risk they faced at each UST site in the Northeast Region and prioritized them for future attention and funding. Golding and Wichman evaluated site contamination by collecting soil gas, soil, and/or groundwater samples from about 500 UST sites in Iowa. They conducted tests in the field with additional testing at laboratories. They also evaluated about 200 of the sites by using a tracer test method.

Once it is clear that a site is contaminated and needs clean up, specific site factors should be addressed to help decide on the method of remediation. Fan and Tafuri present a method that helps to screen the various remediation technologies and find the most appropriate one.

The Symposium Co-Chairmen gratefully acknowledge the efforts of the authors, reviewers, and ASTM personnel that have made this publication possible.

Philip B. Durgin Ph.D.

Veeder-Root Co., Simsbury, CT 06070
Symposium Chairman and Editor

Internal Monitoring

Warren F. Rogers[1]

VOLUMETRIC LEAK DETECTION - A SYSTEMS PERSPECTIVE

REFERENCE: Rogers, W. F., "Volumetric Leak Detection - A Systems Perspective," Leak Detection for Underground Storage Tanks, ASTM STP 1161, Philip B. Durgin and Thomas M. Young, Eds., American Society for Testing and Materials, Philadelphia, 1993.

ABSTRACT: Volumetric testing of USTs has grown exponentially in recent years, stimulated by regulation and by increasing business sensitivity to potential liabilities from leaking tanks. This paper attempts to lay an intellectual groundwork for discussion of such systems and focuses on the objectives of leak detection and the role of measurement precision. The important methodological issues of identifying and discriminating among sources of apparent volume changes which are not related to leakage, and the analysis and interpretation of test results will also be addressed.

KEYWORDS: volumetric leak detection, leak detection, measurement accuracy, precision, leak rate, probability of detection, probability of false alarm, error, statistical inventory analysis, automatic tank gauge

INTRODUCTION

Volumetric testing of underground storage tanks has grown exponentially in recent years, stimulated in part by regulation and in part by increasing business sensitivity to potential liabilities from leaking systems. Research and engineering development of leak detection systems has also expanded dramatically but has been characterized by little evidence of scientific or engineering discipline.
Other than the USEPA's 1988 Evaluation of Leak Detection Methods for Underground Fuel Storage Tanks[1], little has been published which would bring coherence to the topic. The methodologies employed in the USEPA evaluation were reviewed and criticized by Baird.[2] The conclusions drawn in this paper, are based on statistical analysis by the author of underground storage tank manual and automatic tank gauge-generated data over a period of twelve years involving many thousands of samples. Unfortunately, the published material on this subject is largely or totally to be found only in trade literature. The author has been unable to identify anything of relevance in the scholarly literature, hence the paucity of references.
The purpose of this paper, however, is not to present definitive results of documented research but rather to suggest the appropriate questions for research, to identify the variables which are relevant to

---

[1]President, Warren Rogers Associates, Inc., 747 Aquidneck Avenue, Middletown, RI 02840

the objectives of leak detection, and to establish a framework for intelligent debate on the subject.

In short, the author attempts to lay an intellectual groundwork for discussion of such systems. The paper is in three parts. The first part focuses on two issues: the objectives of leak detection, and the role of measurement precision[2]. The second part addresses the issue of identification and discrimination among sources of apparent volume changes which are not related to leakage. The third part deals with the analysis and interpretation of test results.

The technology for detecting leakage from underground storage tank systems is and probably will continue to be dominated by various means of measuring volume changes of the product in the system. This paper is confined to such systems.

Given the wide interest in the volumetric approach, there has been very substantial investment in research and engineering design of the means to accomplish it. This work has tended, however, to be very narrowly focused, and has generally overlooked some fundamental tradeoffs which can lead to more reliable results with substantially lower cost.

As so often happens in the course of technological development, technical goals became confused with system objectives. Obviously the objective of leak detection is to provide a means of identifying and terminating losses of product from tank systems before environmental impairment results. Engineering goals in developing leak detection systems have invariably been stated, however, in terms of leak *rates* to be detected or minimum volumes to be detected within some arbitrary time. As we will demonstrate below, the engineering goal and the environmental objective are fundamentally different. In fact, achievement of the engineering goal has all too frequently defeated the environmental objective.

Regulations pertaining to underground storage tank systems have also, tended to focus on this limited goal. Thus, many bodies of regulation focus entirely on the precision of instruments and the accuracy of measurements which may be taken over very short time frames separated by very extended time intervals rather than on the potential volume of product which may be released into the environment during the hiatus between tests. By mandating extreme short term accuracy, they impose very high unit costs which political realism dictated could only be imposed episodically.

In this article, we examine the process of volumetric leak detection as a parametric system, and suggest criteria for designing systems in a somewhat more balanced way than has hitherto been done.

A Conceptual System

Volumetric leak detection in general consists of three basic activities.
1. Measuring volume change over time
2. Accounting for systematic effects which cause apparent volume changes unrelated to discharge of product from the system
3. Interpreting the measurements

It is necessary, therefore, that the system include:
1. Measurement accuracy
2. Ability to identify and discriminate between physical effects which cause real or apparent volume changes
3. Sensitive, robust means of analyzing and interpreting the measurements.

---

[2] The use of the terms precision and accuracy seems to generate discussion of a somewhat theological nature akin to angels dancing on pin heads. In this paper, the terms are used as follows. "Accuracy" is a quality associated with a measurement actually taken, while "precision" is a quality of the instrument used to make the measurement.

Measurement Accuracy

The pursuit of measurement accuracy through greater refinement of measuring devices has preoccupied research and development activity in this field since its inception. Systems have been developed which are capable, at least theoretically, of detecting volume changes in the thousandths of gallons. Whether such accuracy of measurement is necessary, or for that matter, desirable, has not been questioned.

Overlooked in the pursuit of measurement accuracy were some obvious alternatives and some inherent liabilities induced by approaching the problem of leak detection solely from that perspective. Most prominent among these is the test time duration/ instrument precision trade-off. A leak of product from an undisturbed system will manifest itself even to the crudest possible measuring device if the time over which measurements are taken is sufficiently extended. Over a 90 day period, a loss of .01 gallons per hour (.038 liters/hour) translates to total loss of 21.6 gallons, which is readily detectable from accurate stick measurements in a 1,000 gallon (3785 liter) dormant tank.

In this regard, the pursuit of extreme precision is self defeating. Precision involves costs:; cost of the system and cost of disruption necessary to create and maintain the controlled environment it requires. Cost also, however, mandates constrained testing duration. The more precision, the greater cost and, therefore, the shorter the affordable usage time and, therefore, the need for ever greater precision.

The inherent costliness of extreme precision carries other disadvantages. The major disadvantage from an environmental standpoint is infrequency of use. If the ability to measure ever smaller volume changes is bought at the cost of testing no more than annually, or less frequently, one has to question its contribution to effective and useful leak detection.

And what is the payoff from extreme precision? Presumably, the ability to identify very small leaks in a very short time. But if a leak is very small to begin with and can be detected by simpler and less costly means over more extended time, what has been lost? Certainly not any significant protection of the environment, particularly if the practical consequence of extreme precision is infrequency of testing.

In short, if a system can reliably identify a loss at the mandated tightness testing threshold of .1 gallons per hour (.3785 liters/hr) by extended measurements over thirty days but is sufficiently affordable to permit measurement every 30 days, would that not be preferable to a system which can accomplish the same precision in one hour but can only be employed at intervals of one to two years?

The Time/Precision Tradeoff

To address this issue requires first that we re-define our terms and clearly state our objective. Properly stated, the objective is to limit the amount of product lost from a tank system to some acceptable volume before the loss is identified and terminated. Therefore, the real objective of a leak detection system relates to leak rate detection capability only inferentially. To state the objective in this way is to face up to some painful physical realities. First, it requires explicit acknowledgment that no system is or ever will be perfect. Implementation of any practical leak detection system implies that some residual loss potential will remain.

Regulators, understandably, have been unwilling to declare explicitly that any loss of product is acceptable. The political sensitivities involved in so doing are obvious.

Nonetheless, if one wishes to compare leak detection systems on some basis more meaningful than leak rate detection capability alone, one must consider, in addition, test duration and the duration of time between tests, when no leak detection is being attempted. Such considerations require explicit acknowledgment of product loss. Fortunately, this need not be stated in some fixed gallon amount but can be parameterized.

Also, some care needs to be exercised in defining exactly what is meant by precision. Measuring systems of all kinds have inherent random errors of measurement. The precision of a measuring system is a measure of the inherent errors of the measuring system and the minimum magnitude of the effect to be measured necessary to make it distinguishable from the random fluctuations.

Leak detection systems are subject to two forms of error.
1. Errors which are inherent in the measuring device which are random and irreducible, the noise in the system.
2. Errors induced by unaccounted for variations in the quantity being measured.

The nominal loss rate which the system is capable of detecting and, hence, its nominal precision is really a statement about the first of these, the inherent error rate. The inherent errors can be summarized as a probability distribution and precision as a quantile of that distribution. Thus, to be precise and complete, a description of the leak rate detection capability of a system should be stated, for example, as follows:

With probability of detection of at least .95 and probability of falsely declaring a leak to exist where none does exist of no greater than .05, the system can detect a loss of .1 gallons per hour (.3785 liters).

Two points are worth noting.
1. Precision is a statement about an error distribution. A statement of leak detection capability in terms of leak rate alone without a statement of the probability of error is totally meaningless.
2. The error distribution will vary substantially with varying test conditions. Therefore, the results of a specific test should be stated in terms of the error distribution observed during that test only. Hence, the need for interpretive analysis.

The foregoing considerations lead to this formulation. Let:

g = Total volume of product which is the maximum which may be permitted to escape from the UST system before detection
t_d = The number of days between tests   Then the hypothetical leak detection system should be capable of detecting

$$\frac{g}{t_d \times 24} \quad (1)$$

gallons per hour.

The EPA requires that an acceptable system have probability of detection of at least .95. Simplifying greatly, we could say that this requires that, during the duration of a test, for a leak to be detected, the volume lost must exceed two standard deviations of the noise distribution of the testing system.

Now assume that one candidate test T_1 is completed in one hour and is repeated annually. Another test T_2 is completed over 30 days but is repeated every 30 days. The leak rate to be detected by T_1 is:

$$\frac{g}{365 \times 24} \quad (2)$$

gallons per hour. The duration of the test is one hour. Therefore, the volume loss during the course of the test and that which the test must detect is the same as the hourly leak rate. The required standard deviation required for T_1 is therefore:

$$\sigma_{t_1} \leq \frac{g}{365 \times 24 \times 2} \qquad (3)$$

The leak rate to be detected by T_2 is:

$$\frac{g}{30 \times 24} \qquad (4)$$

gallons per hour. The duration of the test is 30 x 24 hours, and therefore, the standard deviation required for T_2 is

$$\sigma_{t_2} \leq \frac{g}{2} \qquad (5)$$

Therefore, to achieve the same level of environmental protection, a test whose measurement duration is 30 days but is repeated every 30 days may have a measurement standard deviation 8,760 times larger than one required to achieve the same result with an annual test lasting one hour.
Consider now a system at the extreme frontier of the state of the art of precision, capable of detecting .005 gallons per hour (19 ml/hr), in a test lasting one hour. To do so, its error distribution must have a standard deviation no greater than .0025 gallons (9.5 ml/hr). A less sophisticated system, to achieve the same level of environmental protection, testing monthly, with test duration of 30 days, need have an error standard deviation of no better than 21.9 gallons (82.9 liters). A not particularly well trained operator sticking his tank with only moderate care can do this well.
The point here is not to endorse either approach but to place the issue of measurement precision in perspective. Much of the discussion of leak detection systems treats precision as an end unto itself. As the previous discussion should make clear, it is largely meaningless unless placed in the context of test duration, cost and test frequency. In short, as in any engineering application, the costs and benefits need to be analyzed and even a superficial examination shows that the implications for systems and policy would not be trivial.

PART TWO

Systematic Errors

In the previous section, we discussed the inherent measurement accuracy of leak detection systems, their inherent error rates and the implications of these for system accuracy and environmental policy. In this section, we discuss errors which are extraneous to the measuring device but inherent in the quantity being measured. That is, the actual or apparent fluctuations of the volume of product in a tank system which occur during the conduct of a test for leakage.
In an idealized world, the relevant dimensions of an object being measured can be assumed to remain constant during the course of measurement. Furthermore, relevant physical characteristics of the

object being measured are assumed known to a level of accuracy commensurate with the measurement to be attempted.

Neither of these criteria are met when attempting practical field measurement of product volumes in functioning underground storage tank systems. The words "practical" and "functioning" here were carefully chosen. The objective is to test real world systems in their operating environments. Many of the difficulties this imposes can and have been offset under controlled laboratory conditions. The artificialities introduced by such conditions, in attempts to date, have generally led to conclusions which were not supportable or capable of being replicated under field conditions.

It is useful to separate the various error-inducing phenomena into two categories.
1. Those which tend to distort and detract from the usefulness of simple, manual inventory record-keeping as a leak detection mechanism.
2. Those which became evident when, on-site, precision testing methods were introduced, presumably to offset those in the first group.

Historically, this was how the subject evolved. Manual inventory was found to be inadequate as a leak detection system and means were sought to improve on it. Unfortunately, the assumed sources of inventory control deficiencies were not subjected to empirical evaluation. Many of the deficiencies deemed crucial turned out, on subsequent analysis, to be of little consequence. Others, which have subsequently been identified as crucial, were overlooked.

Thus, it was accepted as a given that a dominant contributor to error was the inaccuracy of manual stick measurement as a means of measuring volume. Whence came the overriding concern with precise measurement addressed in the first section. But, with the disruption of operations required for precise measurement came the imperative for severely constrained time duration of measurement. This, in turn, amplified the distorting influence of such variables as temperature fluctuation during the course of the test, thereby introducing a wholly new source of extraneous error which had had little or no impact on inventory control accuracy.

It is one of the ironies of this development sequence that as it became apparent that temperature variation could have a crucial impact on short term precision measurement, it became accepted, again uncritically, that it should have a similar overriding influence on inventory accuracy. Our analyses of inventory records generated from ATG data which incorporated temperature measurements have shown that this is not typically the case.

Now, well after the fact, extensive statistical analysis of many thousands of inventory records, correlated with physical evaluation of the tank systems involved, has definitively established the following:
1. Manual inventory recordkeeping alone is an inadequate means of leak detection in an active tank system for leaks of less than 20 gallons per day.
2. The sources of manual inventory deficiency are correctable, but are not those which the introduction of precision measurement, either testing or in-tank gauging devices, were designed to correct.
3. Measurement (sticking) accuracy is in fact the least critical and most easily corrected source of error in this system. A well trained, conscientious operator can consistently maintain error rates of less than 10 gallons (37.85 liters) on average. As discussed in the previous section, such measurements taken consistently over a sufficient time span can yield measurement precisions as good as, or in excess of, those achievable by a precision measuring device employed over a more limited time.

The inaccuracies and errors which limit manual inventory control as a leak detection system are unrelated to measurement. They are entirely due to the dynamics of tank system operations during the test period. If a tank system can be filled and maintained in an undisturbed state with no additions or removals of product for a period of 15 days or more, daily stick readings, properly executed and analyzed, will provide leak detection accuracy in excess of that achievable by any current precision test method. This has been demonstrated repeatedly and continues to be employed routinely with heating oil systems during their dormant non-heating season.

It is, of course, impractical to take active retail sales or vehicle service tanks out of operation for such extended periods. Therefore, two alternative approaches to leak detection are required.

1. Retain the concept of testing a dormant system but constrain the test period to a practical limit. This is the precision testing approach. As discussed earlier, the requirement for precise measurement escalates sharply as test time is reduced. In addition, however, several additional and very significant sources of error are created by this approach which were never present in inventory records.
2. Retain the concept of testing over extended time but identify and correct for the errors introduced by the dynamics of an active system. This is the approach taken in statistical inventory analysis.

Automatic Tank Gauges

The discussion, to this point, has dealt with the general characteristics and requirements of volumetric testing. The observations and conclusions are general and independent of the specifics of the technical approaches which might be made to implementing them in a practical leak detection device.

From this point onwards, however, we will be somewhat more specific. We will consider two polar opposites in approaching the problem of leak detection in practice; analysis of system inventory in an operating system and analysis of system volume in a system taken out of operation for a brief period, e.g., precision testing.

We need to mention specifically, however, Automatic Tank Gauges (ATG), because they are a hybrid of both approaches. The (ATG) is an automatic gauging system which continuously measures product volume. During active system operations, it can be viewed as an inventory measuring device for an active system. During periods when the system is dormant, e.g., nights and weekends, it can be placed in a "leak detect mode" and, as such, conceptually at least, may be analyzed in the context of precision testing.

In that sense, therefore, with the exception of the issues addressed below, their introduction adds nothing new to the discussion other than the fact that both of the technical approaches addressed are combined in one device. The appropriate parts of the analysis can, therefore, be applied to the relevant modes of ATG operation.

An ATG, however, measures product temperature in addition to product volume. For some inventory analysis applications, this is a valuable, and in some circumstances an essential feature.

As will be discussed below, diurnal temperature variation has a negligable effect on inventory data, provided inventory measurements are taken at the same time each day. For high volume sites, however, with deliveries occurring daily, multiple daily measurements are necessary in order to distinguish leakage from persistent under-delivery effects.

Under these circumstances, diurnal effects become significant, and inventory analysis requires temperature readings coincident with volume measurements.

The leak detect mode, when applied for a single extended time period, is in all ways equivalent to a precision test. Somewhat different is the case when the dormant mode is extended to include non-overlapping intervals when the system is not in operation and the measurements so obtained are combined to produce results equivalent to those derived from a single period.

In this, the ATG test departs very significantly from precision testing practice and interpretation. Each measurement taken involves an error term whose variance is a property of the measuring device. A single dormant mode test requires two such measurements and, hence, twice the inherent measurement error variance. With multiple intervals the variances are additive and, therefore, the sensitivity of the test is progressively degraded. In addition, the individual measurements occur at different product levels. In effect, the device is now addressing a dynamic system subject to problems such dynamics introduce and must account for them.

Extended Time Dynamic System Testing

As noted above, there is no need for any more precision or elaboration beyond simple manual inventory control procedures when a tank is dormant and can be maintained so over an extended period, at least 15 days. When the tank is in active operation, however, several distortions are introduced which must be corrected for.

First, and most crucial, is that the product level is changing constantly throughout the testing period. When testing a dormant system, the height of product in the system is the only relevant variable. When the tank is in active operation, however, volume measurement is required, and this single requirement introduces the following complexities.

1. Varying heights of product in a cylindrical vessel, whose dimensions and orientation in the ground are known only approximately, must be converted to volume.
2. Inaccuracies in the reported volumes of product introduced into the system must be identified and corrected for.
3. Inaccuracies in the volumes of product removed from the system must be identified and corrected for.
4. Changes in total system volume induced by the activity in the system (e.g., temperature differentials between delivered product and product in the system prior to delivery) must be accounted for.

Statistical inventory analysis procedures were developed as a means of addressing these dynamic effects. The methods for doing so will be discussed below.

However, it is useful to pause at this point to consider what can be achieved when this is done. From inventory data alone it can be established:
1. That the tank system is not losing product, through leakage or other means in excess of an amount determined by sticking accuracy and duration of observation.
2. The probability that leakage would have been detected if it existed. This is pre-set to any desired level.
3. The probability that loss of product would be declared where none existed. This is also pre-set.

In short, the system can be tested with any desired probability of detection and false alarm and, if no losses are observed, the degree of accuracy achieved can be exactly quantified. This is a point which is frequently misunderstood. Certification of a leak detection method attests to the performance of that method at one time and under one set of conditions. The environments in which testing in the field is conducted and, likewise, the accuracy with which product volumes are measured for inventory analysis are both highly variable. Therefore, it is necessary, if results are to be meaningful, that data from each individual test conducted be analyzed to determine the leak rate which could have been detected on that specific test. This is an inherent feature of statistical inventory analysis and could be similarly applied in other volumetric test applications.

Adjusting for Volume Dynamics in an Active System

To address the means for overcoming the complexities introduced when moving from a dormant system to a dynamic system, we will first consider, in each case, for illustrative purposes, an idealized approach which is, unfortunately, of limited practicality in real world systems. We will then discuss how this may be adapted in general practice.

Height to Volume Conversion

The ultimate approach in this regard with new or existing systems would obviously be to calibrate each system individually by successively introducing or removing small volumes of product and recording resulting product height. This would eliminate all questions of tank geometry, orientation, dimensions, etc.[3]

In practice, this is very rarely done. Expense is usually cited as the reason. In light of the long term benefits of having close to exact height to volume conversion, the one time expenditure would seem to us more than justified. However, our objective here is to deal with the world as it is, and therefore, we consider alternatives consistent with real world practice.

The next best alternative is to assume that the nominal geometry of the tank reflects its real geometry and measure the dimensions of the tank and its orientation in its excavation. For example, if we are dealing with a flat ended steel tank, we can assume it is a perfect cylinder, measure its length, radius and number of inches of tilt over its length when installed. From those measurements, adjusted as necessary for such details as splash plates, it is a trivial matter to produce a conversion chart or conversion equation for use on a computer or calculator. Minor adjustments are required for such things as hemispherical end plates and tapered walls of fiberglass tanks, but they are also mathematically straightforward.

While the foregoing would seem obvious and sensible steps to take when installing new tanks, they are very rarely taken. Furthermore, most operators are unwilling to bear the expense of uncovering existing tanks to measure them retroactively. Therefore, again to accommodate to the realities of the operating world, we need a third approach.

This is done statistically. The geometry of a cylindrical tank imposes unique characteristics on the numbers generated when height is converted to volume. If the dimensions used for conversion differ in even very small amounts from the dimensions of the tank being gauged, a very distinctive pattern, unique to those dimensional discrepancies, is introduced into the data. This can readily be identified and corrected for by statistical procedures. All that is required is that careful, accurate stick readings be made during a period when tank volume ranges from as close to full to as close to empty as is practical.

We would emphasize that in an ideal world, the preferred approach is the first one described. In fact, there are circumstances where it is the only one which is practical. For example, if the tank shape is severely distorted during installation, flexes significantly in operation or, very critically, if it is deliberately distorted by introducing a lining, then either very exact measurement or gallon by gallon calibration are the only practical alternatives.

Addition and Removal of Product

The preceding extended discussion of the effects of geometry on height to volume conversions was given in detail because the approach is archetypical of the methods used to identify and correct for all other

---

[3] One referee suggested that correlation of ATG readings with dispenser meters would be equivalent. This, of course, assumes that both the ATG and dispenser are performing accurately. But a major function of statistical analysis is to establish whether they are, in fact, functioning properly, and that reasoning is, therefore, circular.

dynamic complexities. Either suitable measurements are made in routine operations to identify and quantify such effects, which is rare in practice, or their potential effects are defined mathematically and their presence and magnitude is derived from the inventory data itself.

Thus, it is well known that deliveries of product are rarely accurate. This could be overcome during a test period by careful metering of delivered product, but this is rarely done. If it is not, we acknowledge that the inaccuracy exists and estimate its magnitude from the data. Fortunately, the appearance of such an event is distinctive. Product volume reported delivered, which is in excess of or less than actual, leaves a permanent imprint of fixed magnitude on all subsequent stick measurements.

Removal discrepancies can be treated similarly. Meters can be calibrated prior to test but, again, rarely are. Instead, the known characteristics of the metering system are defined mathematically. Typically, though not always, metering errors are proportional to volumes dispensed. In this way, they, too, can be identified and removed from the data.

Temperature Effects

As mentioned earlier, temperature variation is typically of less consequence in this context than in precision testing. There are exceptions, however, and these are addressed below.

In precision testing, temperature variation is of consequence because of the extremely short duration of the test. Minor fluctuations which would have little discernable effect on an inventory analysis produce volume variations which can be significantly large relative to the small total volume change the tester needs to identify during the short period of the test.

On the other hand, the system is dormant during the test, so that only short term effects, internal to the tank system, need to be identified.

Inventory analysis on the other hand must deal with a dynamic system over an extended time frame. Three possible sources of temperature variation must, therefore, be addressed.

First, there is diurnal change. If measurements are taken at the same time each day, this is of no consequence. Our analysis of data where multiple readings were taken each day, however, has shown that diurnal effects are significant. So much so that, absent temperature readings, it is not possible to achieve satisfactory results when all readings are merged into a single data set. Data collected at different times must be analyzed separately.

This creates a problem at very active sites where deliveries occur daily. Multiple daily measurements are necessary at such sites to distinguish between delivery discrepancies and other effects. For such applications, temperature measurements generated from an ATG or other source are essential.

The second source is seasonal variation. The Agriculture Department has, for many years, recorded underground temperatures throughout the United States. The data shows that there is typically very little variation, certainly insufficient to affect inventory results over short periods of time on the order of thirty days. There can be, however, significant effects over more extended periods. Therefore, if inventory analysis is attempted over periods in excess of 30 days, underground temperatures should be measured and accounted for.

Third is the effect of introducing product into a tank at a significantly different temperature than the product stored in the tank. This can be of significance when the differences are extreme. However, the effects are less extreme than might be expected due to the dynamics of the system.

For example, suppose that 10,000 gallons of product is delivered into an empty tank where the underground temperature is twenty degrees fahrenheit lower than that of the delivered product. Then it is conceivable that:

$$10,000 \times 20 \times .0007 = 140 gal \qquad (6)$$

of shrinkage could occur.

In order for this to happen, however, it would be necessary for the tank to remain dormant with no additions or removals of product until the temperature decayed to the underground ambient. In this case, the inventory analysis is trivial. Merely record the temperature of the delivered product, the underground temperature, make the appropriate adjustment and measure the product level over time.

In the typical application, of course, this is not realistic, and several dynamic effects enter which greatly mitigate the temperature effect.

First, product is rarely introduced into an empty tank. The initial shrinkage of the delivered product is, therefore, limited to the effects induced by the temperature of the combined products.

Second, product is typically being removed through sales, and, therefore, smaller and smaller volumes are exposed to shrinkage.

Finally, repeated deliveries typically take place before residual product decays to underground ambient. With each addition under these conditions, the temperature of the residual product is higher than it was at the time of the prior delivery resulting in lesser temperature differentials.

Extraneous Errors in Precision Testing

The errors induced by the tank system in the course of precision testing differ substantially and fundamentally from the dynamic effects on inventory of continued system operation. The tank system during precision testing is dormant. The distorting effects are largely functions of the short term duration of the test.

Large volumes of new product are usually introduced into the tank system prior to the test. Consequently, the tanks must be allowed to stabilize to prevent tank deflection, extreme temperature fluctuation, or the formation and dissipation of vapor pockets during the course of the test. All of these effects are known and are routinely referred to in the descriptive materials provided by manufacturers of test equipment. They are large effects particularly when compared to the magnitude of the volume change due to leakage that the system is attempting to detect.

While all of these effects are known, there does not seem to be a consensus as to how they should be measured. Thus, for example, end plate deflection or vapor pockets are acknowledged to exist as distorting influences, but there does not appear to have been any rigorous mathematical treatment of their defining characteristics or, in the case of vapor pockets, their rate of decay and its influences on test results.

Temperature effects are the most frequently discussed, but the discussions have yielded no generally accepted means of measuring them or documentation as to the precision with which they are being measured. There are at least three schools of thought on this.
1. Circulation of product to produce uniformity of temperature.
2. Single thermistor measurement at the mid-point of the tank.
3. Multiple in-tank thermistors with simple temperature averaging.

The Vista research paper[1] addressed this issue, but Baird[2] provides some cogent arguments that the methodologies used were flawed. Absent a more definitive treatment, sufficient to withstand an objective and technically qualified peer review, I believe the issue must remain open.

In view of the clear differences in approach and the lack of consensus on this issue, it would appear that, at the very least, different effects are being measured with very different volume implications. All however, are being interpreted identically. They all assume that the temperature variations they measure, or average, have equal impact on the total volume of product and produce volume changes strictly proportional to it.

For these, if for no other reasons, one has to question whether the claimed and usually demonstrated inherent precision of the measuring devices used, is at all relevant. It is at least arguable and certainly needs to be addressed, whether the precision of such systems is, in fact, determined by the extraneous influences induced by the tank system under test. The measurement of the effects of such influences on the overall precision of the testing systems, has not been well documented and their magnitude is potentially many orders of magnitude larger than errors which are inherent to the measuring system alone and on which its claim to precision is based. Such problems are clearly solvable just as the dynamic influences of system operations have been mathematically analyzed and accounted for. The physics of these various phenomena are straightforward and their mathematical treatment should be no less so.

PART THREE - ANALYSIS AND INTERPRETATION OF RESULTS

In the previous sections of this paper, we dealt with the problems associated with measuring volume changes in a tank system. We identified two distinct and independent sources of error in making such measurements. First are those which are inherent in the measuring device. Secondly, there are those induced by the changing dynamics of the system being measured.

We now address the final step in volumetric testing, the analysis of the measurements derived from the test to determine if the tank system is leaking.

That such post-analysis is not only desirable but essential would seem to be obvious. Leakage is but one of many potential sources of volume change which may take place in a system under test. Frequently, such non-leak related fluctuations will exceed by orders of magnitude the volume loss from leakage which the detection system seeks to identify.

In addition, the inherent random component of error in the measuring system used, whether a mechanical device or manual stick readings will vary from test to test and operator to operator. If specific test results are not subjected to analysis, the magnitude of the random component of measurement is unknown and consequently the magnitude of a leak which could have gone undetected during that test is also unknown.

Certification procedures as mandated by the USEPA serve the useful purpose of establishing that a particular leak detection system, method or apparatus, did in fact function as specified under one set of circumstances at one point in time. They provide no guarantee, nor do I believe they were intended to, that in all subsequent uses, the same standards will be achieved.

Statistical Inventory Analysis, like all other volumetric methods, is based on measurements of volume recorded over time. It overtly and explicitly acknowledges that such measurements include errors inherent in the measuring device, and volume changes which are unrelated to leakage induced by the dynamics of the system being tested.

The accuracy and the limitations of the results achieved on each individual test are calculated and reported in terms of the minimum leak rate which could have been detected in the light of the quality and accuracy of the measurements provided for analysis.

The reason that this issue has been consciously and explicitly addressed from the outset of this methodology was the recognition that in this approach, instrument precision was being traded off for measurement duration.

Until recently, this requirement for post-test analysis was not acknowledged in precision testing methods. I would conjecture that this arose from a conviction that extreme precision could overcome all such problems.

Post-Test Analysis

Post-analysis of test data is required to achieve the following:
1. Identify and correct for volume changes, both real and apparent, which are induced by the dynamics of the system under test but are not leak related.
2. Identify and determine the magnitude of the measurement error introduced by the measuring device during the course of the test. This can have both random and systematic instrument induced components. A common example of the latter are vibratory motions in the balances used in devices which measure buoyancy.
3. Determine if a loss of product exists which is of sufficient magnitude relative to the measurement error to be statistically significant. The level of significance can be pre-set as is done, for example, in USEPA regulations at .05. It can also be a quantity calculated from the test data, what is commonly called a p value.
4. If no significant loss of product is identified, calculate the minimum loss which would have been statistically significant in the presence of the measurement errors generated during the test.

The mathematical and statistical tools necessary to perform such calculations correctly are standard and readily available. Under most circumstances, they would hardly merit further discussion. There are, however, numerous examples from practices in the leak detection industry that suggest that more carefully drawn standards are needed.

An exhaustive list of the problems in this area or a catalogue of the appropriate methods which should be employed is clearly beyond the scope of this article.

Instead, I list some practices which should be avoided.

Examples from Current Practice

1. Automatic tank gauges which use height to volume conversion calculations appropriate to a horizontal tank when the tank is significantly tilted. Some manufacturers assert that this problem can be overcome by positioning the probe at the midpoint of the tank or by averaging results from two symmetrically placed probes. A moment's reflection will show that this assertion would be correct only if tanks had a rectangular cross-section. Such tanks are, unfortunately, somewhat rare.
2. Devices which use linear regression to analyze data which is clearly periodic.
3. Altering levels of statistical significance by recording measurements more frequently.
4. Treating observations which are serially correlated as independent.

However, while flaws of this kind are common wherever test results are subjected to analysis, in many systems currently in use, no analysis of any kind takes places. Furthermore, under current regulations, none is required. Regulations require that analysis of results be undertaken only once in the life of a system, when it is submitted for certification.

This is not to say that the results achieved by these systems are inaccurate or in any way deficient. That cannot be asserted with any more confidence than that they are accurate and effective. In the absence of analysis of the test results, their quality is simply unknown and unknowable. They are unknown to the operator who performed the test, the tank owner who paid for it, and the regulator who required that it be done.

REFERENCES

1. Robert D. Roach, James W. Starr, Joseph W. Maresca, Jr. <u>Evaluation of Volumetric Leak Detection Methods for Underground Fuel Storage Tanks</u>, Risk Reduction Engineering Laboratory, US Environmental Protection Agency, Edison, New Jersey, November 7, 1988.

2. William E. Baird,"Critical evaluation of EPA's UST testing apparatus" <u>Pollution Engineering</u>, pp. 86-89, July, 1988.

Donald W. Fleischer[1]

ERROR SOURCES IN AUTOMATIC TANK GAUGING SYSTEMS

REFERENCE: Fleischer, D. W., "Error Sources in Automatic Tank Gauging Systems," Leak Detection for Underground Storage Tanks, ASTM STP 1161, Philip B. Durgin and Thomas M. Young, Eds., American Society for Testing and Materials, Philadelphia, 1993.

ABSTRACT: There are several sources of error which an automatic tank gauging system must overcome in order to accurately detect and measure a liquid leak from an underground storage tank. Some of these sources are: liquid temperature change measurement inaccuracies, tank chart errors, liquid temperature coefficient of expansion inaccuracies, absolute and differential liquid level measurement inaccuracies, liquid evaporation, tank thermal expansion, liquid expansion in tank piping, vapor pockets, tank deformation and groundwater pressure. This paper describes the occurrence and impact of these errors.

KEY WORDS: automatic tank gauging system, leak detection, error sources, underground storage tank, level measurement, temperature measurement, evaporation, tank chart errors, coefficient of expansion, underground pipes

Automatic Tank Gauging Systems (ATGSs) which meet certain performance criteria can be used to meet EPA requirements as a method of release detection for underground storage tanks (USTs). An ATGS is essentially a device for automatically and accurately measuring the level or depth of liquid in a tank. From this measurement, the volume of liquid in the tank can be determined by cross-referencing the measured level with a depth/volume chart for the tank in question. Besides measuring the liquid level in a tank, most ATGSs also measure the temperature of the liquid and the level of any water at the bottom of the tank.

[1]Director, Technology Development, Veeder-Root Company, Simsbury, CT 06070-2003

18 LEAK DETECTION FOR UNDERGROUND STORAGE TANKS

Since an ATGS can measure liquid volume in a tank, it can also detect leaks above a certain rate, because a leak manifests itself by a change in volume over time. The ATGS does this by performing a leak test during a time when no liquid is being removed or added to the tank. A leak test consists of measuring the liquid levels and temperatures at the start and end of a test, converting the levels to volumes, adjusting for any expansion or contraction of the liquid due to temperature change, then taking the difference of the adjusted volumes and dividing it by the elapsed time of the test to obtain an apparent leak rate. The apparent leak rate is compared to a leak detection threshold. If the apparent leak rate exceeds the leak detection threshold, a leak is declared. Figure 1 shows a typical double-wall UST installation with an ATGS installed. The ATGS probe is located in the center of the tank and a submersible pump connected to underground piping is at the right side of the figure.

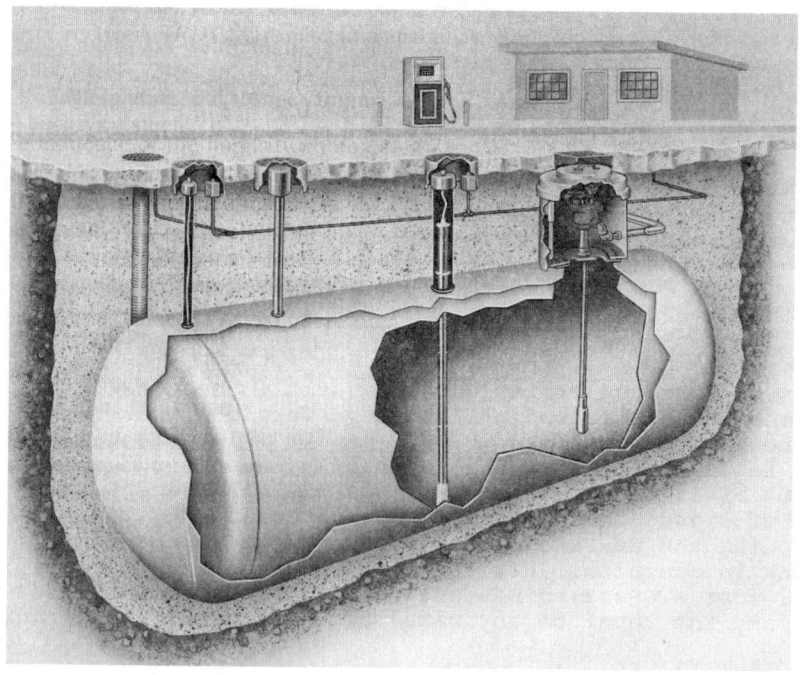

FIG. 1--A typical UST installation with ATGS.

There are several sources of error which can contribute to inaccuracies in ATGS readings. This paper looks at those sources which can affect the ability of an ATGS to accurately measure tank leaks. The findings described here were obtained from measurements made starting in 1978 on several hundred different USTs, representing thousands of leak tests. Well instrumented multiple UST test facilities located in Hartford, CT and Altoona, PA as well as instrumented service stations located in the Hartford, CT area were used to gather precise data on temperature, level, volume, evaporation, tank changes and pipeline effects.

The error sources and their magnitudes cited in this paper apply to USTs of the size normally found at service stations in the U.S., 12,000 gal (45,000L) and below. While larger USTs will experience similar types of errors, their magnitudes generally will be much larger and waiting times or other methods used to reduce their effects will be substantially different than those covered here.

LIQUID TEMPERATURE CHANGE

The effects of expansion and contraction of petroleum products due to temperature changes are well known. In fact, liquid temperature change is the only noise source introduced in evaluating an ATGS under the EPA's Standard Test Procedures. These effects can be calculated if the average liquid temperature change during the test (ΔT), the volume of liquid in the tank (V) and the temperature coefficient of expansion of the liquid ($C_E(T)$) are known. The volume change due to expansion (ΔV_E) is calculated using the following formula: $\Delta V_E = C_E(T) V \Delta T$. There can be errors present in each of the factors, however.

First, determining the exact average temperature change of the liquid in a static tank is an impossible task which would, taken to its ultimate, require an infinite number of temperature sensors. The temperature measurement problem is intensified by the fact that petroleum products have low thermal conductivities. Thus, liquid volumes having substantially different temperatures can exist adjacent to one another with very little heat flow occurring to bring the temperatures into equilibrium. This is especially true in the vertical direction during the summer when soil temperatures are warmer near the ground surface and cooler near the tank bottom. Then, a stable thermal condition can exist where warmer, less dense liquid is at the top and cooler, denser liquid is at the bottom. Under these conditions, vertical temperature differentials of 16°F (8.9°C) have been recorded in partially filled 8 ft (2.4m) diameter tanks.

Most ATGSs attempt to determine the average liquid temperature change from a single vertical array of temperature sensors. There is usually no attempt to determine horizontal temperature differences. Due to the fact that horizontal layers of liquid will tend to have the same density, they will also tend to have the same temperature and therefore, horizontal temperature differences are usually quite small. However, horizontal differences in the temperature of the fill surrounding the tank can cause larger than normal horizontal liquid temperature differences. This can occur if, for instance, there is groundwater flow past one part of the tank or if a portion of the tank is located under a black top surface exposed to the sun.

Substantial stable liquid temperature differences can exist in various parts of the tank without affecting leak test results, providing the average temperature change of the liquid is accurately measured by the ATGS during the leak test. However, it only takes an average temperature change error of 0.03°F (0.017°C) per hour in a full 10,000 gal (38,000L) tank of gasoline to produce a thermal volume change of 0.2 gal/h (0.76 L/h), the leak rate an ATGS must detect to comply with EPA regulations. Such an error rate, however, cannot be tolerated, since a 0.1 gal/h (0.38 L/h) threshold is typically used for declaring a 0.2 gal/h (0.76 L/h) leak. Therefore, the total of all leak measurement errors must be kept to less than 0.1 gal/h (0.38 L/h). In order to have a low probability that the total of all leak measurement errors will exceed this limit, it is assumed here that any single error should be kept to less than 0.03 gal/h (0.11 L/h). Therefore, this would mean that the average temperature change error of the gasoline in the 10,000 gal tank should be kept to less than 0.0045°F/h (0.0025°C/h). This was calculated using a volumetric coefficient of expansion of 0.00068 gal change/gal stored per °F (0.00122 per °C) for gasoline.

Obtaining average liquid temperature change errors of less than 0.0045°F/h (0.0025°C/h) requires substantial waiting times during which no fluid is added to or removed from the tank for any temperature changes to stabilize. ATGS protocols typically impose waiting times of several hours between any product additions to the tank and the start of a leak test. However, a waiting time of at least one hour should also be observed between the end of dispensing and the start of a leak test to reduce the error causing effects of temperature changes that occur throughout the tank which are not measured due to their distance from the temperature sensors. Extremely active tanks require longer waiting times.

Knowing the exact volume of liquid stored in the tank is

not as critical a factor as knowing the exact temperature change, but an error in measuring the volume can have an effect on determining the thermal volume change (volume change considering only the effects of temperature). There are many factors which can cause an error in measuring the volume of fuel in a tank, but probably the worst source of error is in the tank charts. A volume error of 5% in constructing tanks, especially steel tanks, is accepted by industry and occurs frequently. Such an error coupled with a 0.09°F/h (0.05° C/h) temperature change in a full 10,000 gal (38,000L) tank of gasoline would cause an error of 0.03 gal/h (0.11 L/h) in determining the thermal volume change. This error would exist even if the ATGS accurately measured the average liquid temperature change. Therefore, sufficient waiting times should be allowed for the average temperature change of the liquid to not just stabilize, but to approach zero.

Tank chart error can be minimized by calibrating the tank. This can be accomplished by performing a metered fill, during which measured quantities of liquid are added to the tank and the corresponding increases in liquid height are recorded. Another way is to keep records of amounts of fuel dispensed from the tank and the corresponding decreases in liquid height that they cause. From either set of data a more accurate tank chart can be calculated.

The coefficient of thermal expansion of the liquid is not measured by the ATGS. It is a manual measurement typically performed by the petroleum refiner. Therefore, an average value for the liquid is usually stored in the ATGS control unit and used for the calculations. The source of the error is in the fact that there can be differences in the coefficient for a particular type of petroleum product from one manufacturer to another, from one season to another and even from one refinery run to the next for the same manufacturer. Some ATGSs have the capability for adjusting the coefficient used for the liquid contained in the tank, but for this feature to be useful would require the measurement of the coefficient of the liquid in the tank after each delivery.

With fuels constantly being reformulated to achieve cleaner burning, it can be expected that the coefficients of expansion of petroleum products will be changing also. A small sampling of the coefficients currently being used by ATGSs to calculate thermal volume changes in gasoline shows a range of 0.0006 to 0.0007 per °F (0.00108 to 0.00126 per °C). This would mean that for a given tank of gasoline, one of the ATGSs would have a coefficient error of at least 0.00005 per °F (0.00009 per °C). Such an error, coupled with a 0.06°F /h (0.033°C/h) temperature change in a full 10,000 gal (38,000L) tank would cause an error of 0.03 gal/h (0.11

L/h) in determining the thermal volume change. Again, this error would exist even if the ATGS accurately measured the liquid temperature change. Once again, this demonstrates the importance of allowing sufficient waiting times for the average liquid temperature change to approach zero.

LIQUID HEIGHT MEASUREMENT ACCURACY

Another source of leak rate error is differential fluid height measurement accuracy. This is the accuracy of the measurement of the change in height of the fluid in a tank due to a leak or thermal expansion. Again, using a 10,000 gal (38,000L) tank with an 8 ft (2.4m) diameter, but this time half full, an error of 0.00045 in. (0.00114 cm) during a 2 hour test would cause a leak rate error of 0.03 gal/h (0.11 L/h). Although 0.00045 in. (0.00114 cm) is a rather small error, generally speaking, the better ATGSs have differential height accuracies better than this.

The error related to differential height measurement occurs in the conversion of that measurement to a volume, which is required to determine the rate of any leak. Essentially, the conversion requires the determination of the cross sectional area of the liquid surface in the tank. This is then multiplied by the height change during the leak test to arrive at the volumetric change. Several factors can contribute to errors in this calculation.

The first factor of error in determining liquid surface cross sectional area is in the tank chart. As mentioned previously, significant errors in these charts are not uncommon. In calculating the volumetric change during a leak test, a tank chart that is in error by 5% would cause a 5% error in the calculation.

The second factor of error in determining liquid surface cross sectional area is the absolute height accuracy of the tank gauge. This is usually pretty good for most high grade tank gauges, typically being within 0.1 in. (0.25 cm). This magnitude of height error is not enough to cause a significant volumetric change error. However, liquid height is usually measured from the bottom of the tank gauge probe and determining if the probe was installed such that it accurately indicates the depth to the tank bottom is difficult.

If there are several openings in the top of an UST and a tank gauge probe is inserted in each one, the liquid height readings will typically vary between the openings by as much as a few inches. This can be caused by the bottom of the probe resting on some accumulated dirt or rust scale or on a dent or other distortion of the tank bottom. A 10,000 gal (38,000L) tank with an 8 ft (2.4m) diameter is approximately

27 ft (8.2 m) long. When filled with gasoline, it contains approximately 30 tons (27.2t) of liquid. This amount of load will cause the bottoms of most USTs, especially fiberglass ones, to become distorted along their length. Also, the probe may not have been installed in a perfectly vertical orientation.

A more common cause of error in liquid height measurement accuracy is tank tilt. No UST bottom is perfectly horizontal and in most cases the tank is purposely tilted, usually with the fill pipe end a few inches lower so that any water on the bottom will accumulate near the end that allows it to be removed through the fill pipe. The tank tilt effect can be minimized by locating the tank gauge probe in the middle of the UST or by programming a tilt offset factor into the tank gauge control unit, which causes the tank gauge level reading to be the same as if the probe were located in the middle of the UST. However, the nominal tank diameter may not be the same as that stored in the ATGS. Welded steel USTs can have diameters different than on their tank charts, due to manufacturing variances. In addition, fiberglass USTs have diameters which purposely vary from the ends to the middle. Also, as USTs are filled with liquid, they tend to flatten, reducing the vertical diameter compared to when empty.

When all of these factors are considered, even carefully installed probes can have their bottoms located such that their height readings are in error by an inch (2.54 cm) or more versus the tank chart.

In a 96 in. (240 cm) diameter UST filled with fluid to the 90 in. (229 cm) level, a 1 in. (2.54 cm) error in the absolute height reading would cause an 8.2% error in the volumetric change calculation. The effect of the absolute height error on the volumetric change calculation ranges from virtually zero error in a half-filled tank to thousands of percent in a filled tank. Whereas the percentage effect of the chart error is independent of the fluid height.

If the volumetric change during a leak test were solely due to a 0.2 gal/h (0.76 L/h) leak with no thermal effects, the consequences of chart error and height measurement error would not be great. A 5% tank chart error would cause a leak rate error of 0.01 gal/h (0.04 L/h) and the 1 in. (2.54 cm) absolute height error would cause a 0.016 gal/h (0.061 L/h) leak rate error. However, if the volumetric change were due to thermal volume changes with or without a leak, the errors could be much more dramatic. In a 96 in. (240 cm) diameter 10,000 gal (38,000L) UST filled to the 90 in. (229 cm) level with gasoline experiencing a 0.1°F (0.056°C) temperature change per hour during the test, the 5% tank chart error would cause a 0.033 gal/h (0.12 L/h) volumetric change calculation error. The 1 in. absolute height error would cause a 0.054 gal/h (0.2 L/h) volumetric change calculation error.

As the liquid level in the UST approaches the top, a 1 in. height error takes on even worse consequences. If the liquid level is at 95 in. (241.3 cm) in the tank example used above, but the ATGS indicates 94 in. (238.8 cm), a 0.1°F (0.056°C) per hour liquid temperature change would cause a 0.28 gal/h (1.06 L/h) volumetric change calculation error. If the UST was filled, so that liquid was 0.25 in. (0.64 cm) up in the riser pipes, a 1 in. (2.54 cm) height error, which resulted in a 95.25 in. (241.9 cm) level reading instead, would result in a much greater volumetric change calculation error. Assuming the UST was level, with a circular cross section and three 4 in. (10cm) and one 2 in. (5cm) diameter riser pipes, a 0.1°F (0.056°C) per hour liquid temperature change would cause an 89 gal/h (337 L/h) error, because at 95.25 in. the surface area of the tank would be used in the volume calculations rather than the surface area of the riser pipes which should be used at 96.25 in. Even with zero temperature change, a 0.2 gal/h (0.76 L/h) leak would be misinterpreted as a 26 gal/h (98 L/h) leak. In this case, height errors much less than an inch (2.54 cm) still would result in unacceptable errors.

The large volumetric change calculation errors caused by even modestly incorrect absolute height readings, especially when coupled with liquid temperature changes, show that leak tests should not be performed using an ATGS on filled or almost filled USTs. In general, leak tests using an ATGS should be limited to USTs containing 95% or less of their full volumes.

LIQUID EVAPORATION

Evaporation can be a troublesome source of error, particularly for gasoline storage tanks. It can cause a reduction in liquid height which is indiscernible from a leak. The rate of fluid loss due to evaporation is affected, among other factors, by the surface area of the gasoline in the tank as well as its Reid Vapor Pressure (RVP) and temperature. It is also affected by the temperature and hydrocarbon saturation level of the air above the liquid.

The RVP of gasoline varies with season and can also vary with grade of fuel. RVPs are higher in the winter months to offset the effect of lower temperatures on vehicle operation. The surface area of the gasoline will vary with the size of the tank and with how full the tank is. A nearly full UST of the typical horizontal cylinder shape will have much less surface area than a half full tank.

The rate of exchange of the air above the liquid with that outside the tank is also a factor. An active tank will have vapor exhausted from the vent pipe when the tank is filled and have fresh, non-saturated air drawn in as fuel is dispensed. Even static tanks can exchange air through vent pipe breathing as vapor in the tank thermally expands and contracts or if wind passes over the vent pipe opening.

Suppliers of gasoline who have analyzed vapor losses generally consider that, on average, 0.14% of the liquid gasoline which passes through an UST in a typical service station is lost to evaporation. Most of this loss occurs during dispensing operations as fresh air is drawn into the tank through the vent pipe to replace the dispensed fuel. Net evaporation then takes place until the vapor levels in this fresh air reach the point where evaporation and condensation are at an equilibrium. The vapors are then expelled from the tank during the next delivery. Thus, a full 10,000 gal (38,000L) tank would lose approximately 14 gal (53L) of gasoline through evaporation if it were slowly emptied by dispensing.

All other factors being equal, the rate of evaporation is roughly proportional to the rate of dispensing. If 5,000 gal (19,000L) of gasoline are dispensed from a tank in one day, the liquid that is lost to evaporation will evaporate during dispensing and at a diminishing rate after dispensing. In the example given, the evaporation rate could easily exceed 0.2 gal/h (0.76 L/h) of liquid during much of this time.

Unless it is known that the ATGS attempts to compensate for vapor loss during a leak test, a leak test should not be started for several hours after the cessation of dispensing in a high volume service station. In stations which have Stage II vapor recovery systems, evaporation losses are substantially reduced and waiting times can be shorter. Usually an hour will suffice.

TANK TEMPERATURE CHANGE

While the effects of temperature changes on the expansion and contraction of the liquid in the tank are usually factored into the results of ATGS leak tests, the effects on the expansion and contraction of the tank itself are not. This thermally induced error is much less than that caused by the liquid, but it can still be significant. Moreover, the magnitude of the error cannot be determined using an ATGS alone.

An ATGS generally measures temperature changes of the

liquid in the center of the tank or along a vertical line in the tank. In any event, the temperature of the tank itself is not measured. The tank temperature is determined more by the fill surrounding it than by the tank contents, especially if the tank is partially submerged in groundwater. Earth has about ten times the thermal conductivity of petroleum, 1.67W/(m·K) for earth and 0.162 for petroleum.

The linear expansion of steel is approximately 7.5×10^{-6} per °F (13.5×10^{-6} per °C). For a 1°F (0.56°C) change in tank temperature, this would result in a volumetric change of 0.23 gal (0.87L) in a 10,000 gal (38,000L) tank. The expansion coefficients of epoxy used in fiberglass tanks are larger than for steel, but the actual tank expansion is determined more by the fiberglass reinforcement and by the direction of filament winding. These vary by manufacturer, so determining an actual value of expansion for a fiberglass tank is difficult. Fiberglass itself, however, has a coefficient of expansion close to that of steel. The leak rate error due to tank thermal changes for a two hour leak test will exceed 0.03 gal/h (0.11 L/h) for a 0.3°F (0.17°C) tank temperature change. This emphasizes the importance of waiting a sufficient time for temperatures inside and outside the tank to stabilize before performing a leak test.

PIPELINE TEMPERATURE CHANGE

An often overlooked source of error in UST leak tests exists in thermal changes of fuel in the pipe lines running from the UST to the dispensers. During much of the year, the pipes, which are usually closer to the surface of the ground than the UST, are at a different temperature than the UST and its contents. In the winter months, the pipes can be much colder than the fuel in the UST and in the summer much warmer.

The source for error arises when a tank leak test is started after dispensing before the temperature of the fuel in the pipes has had time to stabilize. Most fuel dispensing systems in the U.S. use submersible pumps in the USTs to pump fuel through the pipes. These pumps contain a check valve to prevent drainage of the pipes back into the tank during times when the pump is off. However, if this check valve functioned as a normal check valve, in the summer months the cooler fuel pumped through the warmer pipes would expand when the pump was shut off, causing tremendous increases in pressure inside the pipes. To prevent this pressure build-up in the pipes, the pumps contain a functional element which limits the pump-off pipe pressure to approximately 12 psi (82,740 Pa). Any fuel expansion which takes place and would normally increase pipe pressure above this amount, is drained back into the tank.

The amount of fuel which drains back into the tank depends upon the type of fuel, the volume of the pipeline and the temperature increase of the fuel in the pipeline. The temperature increase can be substantial. It would not be unusual for piping located under a black top surface in the summer to be at 100°F (37.8°C). If fuel in the UST were at 75°F (23.9°C), this would produce a temperature difference of 25°F (14°C). A typical service station might contain 150 feet (45m) of 2 inch (5 cm) ID piping to connect an UST to several dispensers. Larger stations can have double this amount. If this 150 feet (45m) of pipe contained gasoline which increased in temperature by 25°F (14°C), the gasoline would expand by 0.42 gal (1.6L), all of which would return to the UST, because the line pressure would remain constant at the functional element relief pressure. Even if the temperature changed by only a modest 10°F (5.6°C), which would be quite normal, 0.17 gal (0.64L) would return to the tank.

If a tank is located at a truck stop, which typically has larger and longer pipelines, the problem becomes much worse. A 25°F (14°C) rise in a 500 foot (150m), 3 inch (7.6 cm) ID pipeline would return 3.1 gal (11.7L) to the tank and a 10°F (5.6°C) rise would return 1.2 gal (4.5L).

The effect of fuel returning to the tank from the pipes during a leak test would be to mask any leak which is present. The opposite effect can take place in the winter when warmer fuel is pumped into colder pipes. Then the fuel in the pipeline can contract to the point where the pressure drops to less than zero and sucks fuel from the tank into the pipes. Since the pipeline would contract slightly with the reduction in pressure, the amounts of fuel transferred would be slightly less than in the summer for any given absolute value of temperature change. Fuel sucked from the tank into the pipeline in the winter would appear as a leak during a leak test.

Temperature changes in the fuel in a single-wall pipeline take place very rapidly, most of the change taking place in the first hour. Therefore, at least an hour should be allowed after dispensing before starting a leak test. Longer times should be allowed if ambient temperature extremes exist or if long pipelines are involved. Heat transfer from double-wall flexible or fiberglass piping takes longer than for single-wall piping, but at a slower rate. A longer wait time would be prudent for facilities equipped with such piping.

MISCELLANEOUS ERROR SOURCES

Vapor Pockets

As mentioned before, performing a tank leak test with an ATGS on an UST which is full or almost full should be avoided. Besides the large leak rate errors caused by small absolute height measurement errors, another problem which can occur is the formation of vapor pockets. These pockets can form when vapor is trapped by the high level liquid at the high end of a non-level tank and in the free spaces of the capped fill pipe, ATGS probe riser pipe and submersible pump riser pipe. Temperature changes of the vapor during a leak test can cause it to expand or contract, which in turn can cause non-leak related changes in the liquid level.

Tank Deformation

Another problem related to full or almost full USTs is structural deformation. As the tank is filled, the increased load tends to make the tank bulge. The problem is that this bulging does not take place instantaneously, but gradually over a period of hours. During this bulging period, as the internal dimensions of the tank are increasing slightly, the liquid level is changing. Likewise, after a rapid, substantial decline in liquid level, the internal dimensions of the tank will tend to decrease slightly over a period of time. If a tank leak test is performed during either of these times, the results will be in error.

The structural deformation of an UST which is caused by changing liquid levels eventually stabilizes. Therefore, waiting times of at least a few hours should be observed after rapid, substantial liquid level changes in either direction, before initiating tank leak tests. If the level change occurs gradually over a several hour period, as usually happens during normal dispensing, the deformation also occurs gradually and a waiting time for this effect would not be necessary.

Groundwater Pressure

An ATGS may not detect a leak in an UST which has a hole in it if, at the time of the test, the hole is located beneath the surface of groundwater surrounding the UST and the groundwater head at the hole is approximately equal to the internal head of liquid at the hole. In this case, no substantial amount of liquid or water is either leaving or entering the tank. However, this static equilibrium is not likely to last for long and either the groundwater level or internal level will eventually change. Then, the next time a tank leak test is performed, a net outflow or inflow of liquid will be measured by the ATGS. If a net inflow of liquid occurs due to the groundwater head being greater than the internal head, besides registering a negative leak, the ATGS may also register an increase in tank bottom water.

CONCLUSIONS

An ATGS can be a versatile and convenient device for automatically keeping track of liquid inventory in USTs and for detecting losses of this inventory through leaks. This can protect the environment and prevent the necessity of costly cleanups. However, in order to accurately measure and detect any leaks which develop, the ATGS must overcome the effects of several sources of error which can influence its results. To reduce these effects the ATGS user should understand how each of these errors can occur and how they can be minimized. The user should also ask the ATGS manufacturer what steps they have taken to control the effects of these errors.

Sources of error include liquid temperature change inaccuracies, absolute and differential liquid level measurement inaccuracies, tank chart errors, liquid temperature coefficient of expansion inaccuracies, liquid evaporation, tank thermal expansion, pipeline liquid thermal expansion, vapor pockets, tank deformation, and groundwater pressure. These errors can either mask or falsely indicate a leak by indicating changes in the height or volume of liquid in the tank.

REFERENCES

[1] U.S. Environmental Protection Agency, "Part 280 - Technical Standards and Corrective Action Requirements for Owners and Operators of Underground Storage Tanks," Washington, D.C.: Federal Register/Vol. 53, No. 185, September 1988.

[2] National Fire Protection Association, "Recommended Practices for Underground Leakage of Flammable and Combustible Liquids" (NFPA 329), Quincy, MA: National Fire Protection Association, December 1986.

[3] Schwendeman, T.G. and Wilcox, H.K., "Underground Storage Systems," Lewis Publishers, Chelsea, Michigan, 1987.

[4] Roach, R.D., Starr, J.W. and Maresca, J.W., Jr., "Evaluation of Volumetric Leak Detection Methods for Underground Fuel Storage Tanks," Cincinnati, Ohio: U.S. Environmental Protection Agency, November 1988.

[5] Flora, J.D., Jr. and Bauer K.M., "Standard Test Procedures for Evaluating Leak Detection Methods: Automatic Tank Gauging Systems," Washington, D.C.: U.S. Environmental Protection Agency, February 1990.

Jairus. D. Flora, Jr.[1], William. D. Glauz[1], G. Joe. Hennon[1]

LEAK DETECTION METHODS FOR AIRPORT HYDRANT SYSTEMS

REFERENCE: Flora, J. D., Jr., Glauz, W. D., and Hennon, G. J., "Leak Detection Methods for Airport Hydrant Systems," Leak Detection for Underground Storage Tanks, ASTM STP 1161, Philip B. Durgin and Thomas M. Young, Eds., American Society for Testing and Materials, Philadelphia, 1993.

ABSTRACT: Airport hydrant systems are large underground pressurized piping systems. Typically, these systems operate with a pressure of about 150 psi and range in diameter from 6 in. to 16 in.. Systems vary in their design, but typically have trunk lines and laterals that end in hydrant pits. These systems are orders of magnitude larger than piping systems found at retail fuel outlets. The lines themselves may contain 250 000 to 500 000 gal of product. Since they are large and operate at high pressure, the potential for environmental damage in the event of a leak is correspondingly increased.

Airport hydrant systems are currently deferred from the leak detection requirements of the federal EPA regulations covering underground storage tanks. However, several states, eg. Texas, have neither exempted nor deferred these systems. These states are regulating hydrant systems on a case by case basis. There is a clear need for leak detection methods for such systems.

A number of approaches to leak detection for hydrant systems have been proposed. These include double wall piping with interstitial monitoring, external monitoring methods for vapor or liquid product on the water table, tracer methods, pressure-step methods, and inventory methods. This paper describes several approaches to leak detection for hydrant systems. It also provides estimates of the performance that may be expected from each method. It relates these performance estimates to the EPA standards for nonexempt/nondeferred systems. The approximate cost to install and operate the methods is discussed.

KEYWORDS: pipeline, leak detection, hydrant system, statistical evaluation

---

INTRODUCTION

The EPA UST regulations (40 CFR Part 280) specify a number of requirements for release detection for underground storage tanks (USTs) and piping. These release detection requirements are documented in Subpart D (paragraph 280.40 to paragraph 280.50). Airport fueling hydrant systems are presently deferred from these federal release detection regulations; however, several states have neither exempted nor deferred these systems. These states are regulating hydrant systems on a case by case basis. This paper describes several approaches to leak

---

[1]Senior Advisor, Principal Advisor, and Principal Chemist, respectively, Midwest Research Institute, 425 Volker Boulevard, Kansas City, MO 64110

detection for hydrant systems, estimating the performance that may be achieved by each method and providing qualitative cost estimates.

Airport hydrant systems are large underground pressurized piping systems. A typical pressurized piping system at a retail service station may be 2 in. (5.1cm) in diameter, 200 ft (61 m) long, and operate at a pressure of about 30 psi (210 kPa). In contrast, an airport hydrant system may have pipes ranging from 6 to 18 in. (15 to 45cm) in diameter, have a total length of 10 miles (16 km), and operate at a pressure of 150 psi (1000 kPa). The lines themselves may contain 250 000 or more gal (960 000 l) of product. The hydrant systems are usually supplied with product stored in aboveground storage tanks.

There are approximately 40 hydrant systems at commercial airports in the United States. Several overseas airports have hydrant systems, and the U. S. military operates a large number of them. Hydrant systems at commercial airports are of unique design, while there are four standard designs for military installations. To provide a standard for comparison, a hypothetical hydrant system is later defined and used to obtain comparative performance and relative cost data.

The federal EPA regulations contain requirements for both tanks and their ancillary piping. It is not obvious whether a hydrant system is more like a tank or its piping. It is more like a tank from the standpoint of the volume of product it contains. From a pressure standpoint, however, a hydrant system is more like pressurized piping than a tank, since USTs are essentially unpressurized. Clearly, the size and pressure in a hydrant system pose the potential for environmental damage several orders of magnitude greater than that of retail service station lines.

LEAK DETECTION CONSIDERATIONS FOR HYDRANT SYSTEMS

EPA regulations for USTs provide a number of acceptable methods of leak detection. The intent of the regulations was to specify standards for protecting the environment and to leave the technology to meet the standards to industry. The EPA does not recommend one method over another, provided that the method chosen for leak detection meets the relevant performance standards. In general, the performance standards specified by the EPA require detection of a specified leak rate with a probability of [at least] 95% and a probability of false alarm of [no more than] 5%.

The EPA recognizes that different leak detection methods provide different degrees of protection and assurance against environmental contamination. There is a tradeoff between the size of a leak a method can detect and the frequency with which the test is applied. A large leak should be detected shortly after it begins. However, a small leak can still cause substantial environmental damage over time, so it should also be detected, but at less frequent intervals. For example, the EPA regulations for pressurized piping require a leak detection method that can detect a leak equivalent to 3 gal/h (11.4 L/h) at 10 psi (69 kPa) within one hour of its occurrence. A less frequent test that can detect smaller leaks is also required in conjunction with this. One option is an annual test of piping with a method that can detect a leak of 0.1 gal/h (0.38 L/h) at 1.5 times the operating pressure. For most retail service stations this translates to a test at 40 psi (275 kPa). A second option for piping is a monthly test capable of detecting a leak of 0.2 gal/h (0.76 L/h) at operating pressure. Other standards are applicable for monitoring the interstitial space of a double wall pipe or a pipe with secondary containment, or for external monitoring (vapor or groundwater) of the pipe.

Because of the size of hydrant systems, there are two distinct aspects of leak detection. One is the identification of a leak, while the other is the location of a leak if one exists. It is not feasible to excavate the entire hydrant system to find or fix a leak; rather a leak must be pinpointed with some accuracy so that it can be corrected rapidly and at reasonable cost. Ideally, a leak detection method for hydrant systems should accomplish both functions--identify the presence of a leak and determine its location.

Of the methods listed in the EPA regulations, several have potential applicability to hydrant systems. For example, various electronic or mechanical devices are available to meet the 3 gal/h (11.4 L/h) at 10 psi (69 kPa) requirement for smaller lines. This leak rate is roughly equivalent to a 45 gal/h (170 L/h) rate at the typical hydrant system operating pressure of 150 psi (1035 kPa). These might be adapted to hydrant systems. Tightness testing of the system, either with a permanently installed module or with separate equipment might be used. Vapor monitoring of the soil around the hydrant system is possible for some sites as is groundwater monitoring. Interstitial monitoring would be possible for double walled or secondarily contained systems.

CHARACTERISTICS OF HYDRANT SYSTEMS

General Characteristics--Hydrant systems consist of large diameter pipes as trunk lines carrying fuel from the tanks and smaller laterals to each hydrant pit. A number of valves are included in the system to allow portions of the system to be isolated for maintenance. The hydrants have valves to connect them to the hydrant truck and planes for fueling.

Size--The size (volumetric capacity of the pressurized portion) of hydrant systems is at least an order of magnitude larger than typical underground storage tanks; capacities of several hundred thousand to one million gal are common. Thus, a leak of a given amount, such as the EPA performance requirement of 0.1 gal/h for tank tightness test methods, represents a proportionally smaller percentage of a hydrant system capacity as compared to the typical underground tank capacity.

Pressure--Hydrant systems are typically operated at pressures of 150 psi (1040 kPa) or more, whereas underground storage tanks are at atmospheric pressure. The weight of the stored product results in only a slight increase above atmospheric pressure at the bottom of the tank. Therefore, a hole in the system of a size that would produce a leak of 0.1 gal/h (0.38 L/h) if it were in a tank, would leak at an increased rate of perhaps 3-5 gal/h (11-19 L/h) under hydrant system pressure.

Vacuum--Hydrant systems cannot realistically be subjected to a vacuum due to the design of the hydrant valves, which would unseat if the line pressure became less than atmospheric. Therefore, leak detection devices employing vacuum techniques would not be applicable without considerable modification to the hydrant system.

Fuel Compressibility and Steel Elasticity--In conjunction with the high pressures and large sizes, the compressibility of the aviation fuel and the elasticity of the steel piping system are important. For example, in a 300 000 gal (114 000 l) system filled to capacity, about 330 additional gal (1250 l) of Jet A fuel would have to be pumped into the system to bring it from atmospheric pressure to an operating pressure of 150 psi(1080 kPa). Or, in terms of leak detection, a leak of 2.2 gal (8.3l) from the system would only drop the pressure from 150 psi to 149 psi (1040 to 1028 kPa).

Thermal Expansion and Contraction--Jet A fuel, like any material, expands and contracts with temperature changes. This is a most critical aspect of airport hydrant systems, and the way that thermal effects interfere with several leak/detection methods must be well understood. Considering only the fuel (its expansion or contraction is of greater impact than that of the piping, but both must be accounted for), a 1-degree F (5/9°C) drop in temperature is equivalent to a volume change of about 150 gal in a 300 000-gal system (570 l in a 114 000 l system). In other words, a loss (leak) of 0.1 gal/h (0.38 L/h) would be exactly compensated for by a gain in volume if the temperature of the fuel increased 0.0007°F (0.0004°C) per hour.

Test Period--Service station-oriented leak detection methods for underground storage tanks generally require that the tested tank or line be taken out of service during the test period. For many methods, this period may include an overnight stabilization time, so the system may be out of service for 8 to 12 hours. Hydrant systems at commercial airports are typically in operation 20 to 22 hours a day. Thus, it is not technically and economically feasible to perform testing except during the short period of inactivity.

Valve Weepage--Hydrant systems include numerous valves to enable isolation or interconnection of various segments of the system and, of course, the many hydrant assemblies (typically, two per airline gate). Each hydrant assembly, located in a hydrant pit, contains a pressure control valve and a shutoff valve for maintenance. These various valves may intermittently seep or weep (under a surge of high pressure, for example). The seepage is contained and redirected to another part of the system, or is manually transferred to a waste oil system as part of daily pit servicing. Since the extent of this seepage affects the level of the leak rate that can be distinguished, maintenance practices should be utilized that minimize leakage.

Leak Detection Limits--The EPA UST regulations specify leak detection limits in terms of gal/h, independent of the size of the tank. This is a reasonable approach for these smaller and unpressurized tanks, given that the objective is to protect the environment and that a given amount of contamination is a problem independent of the size of tank from which it came. It also presents technically achievable measurement standards, since underground storage tanks are usually in the 8 000- to 10 000-gal (30 000 l to 40 000 l) capacity range, with some perhaps as large as 50 000 gal (200 000 l).

For large pressurized systems, like airport hydrant systems, the laws of physics dictate that leak detection limitations are not in terms of absolute leak rates, but of leak rates relative to the capacity and pressure of the system. As discussed above, a hole in an unpressurized tank that might produce a release of 0.1 gal/h (0.38 L/h) would produce a release an order of magnitude or more greater in a system pressurized to 150 psi (1080 kPa). And, increasing the size of the system from 10 000 gal (40 000 l) to 1/4 million gal (950 000 l) requires that the detection technology be more than an order of magnitude more accurate.

Investigation of the question of leak detection limits for hydrant systems was begun recently in Europe as well. At the October 1989 international conference, "Modern Practice in Handling Aviation Fuel at Airports," presented by the Institute of Petroleum, in London, one of the key speakers discussed "Hydrant System Integrity Monitoring". He reported that the Institute of Petroleum Working Group had adopted as a guideline an upper limit of detection acceptability of 0.005% of system capacity per hour at the system operating pressure. For a 300 000-gal (1 140 000-l) system this corresponds to a leak rate of 15 gal/h (56.8 L/h). As the remainder of this paper will indicate, we believe that this goal can be met by several release detection methods, and perhaps improved upon. However, these improvements will not be likely to meet the EPA underground storage tank leak detection criteria.

Hypothetical System--Since hydrant systems vary considerably, a hypothetical hydrant system was defined to provide a common basis for comparison. The hypothetical system is for a 20-gate concourse with two hydrant pits per gate. The pipe from the tanks was 18 in. (45.7cm) in diameter and 10 000 ft (3.05 km) long. The system included a 3 000 foot (914.4m) loop of 14 in (35.6cm) diameter pipe and laterals of 6 in. (15.2cm) diameter with a total length of the laterals of 800 ft (244m).

Steel pipe was used in the model. Using the properties of steel and of the assumed product, Jet A, the bulk modulus of the system, $B = dP/(dV/V)$, was about 130 000 psi (897 000 kPa). The total capacity of the system was about 157 000 gal (594 000 l). The assumed daily throughput averages 250 000 gal (946 000 l) with a peak flow rate of 5250 gal/min (19 900 l/min). The system is assumed to be connected to storage capacity of 1 000 000 gal (3 785 000 l) in four aboveground storage tanks of equal size.

The approximate cost (in 1990 dollars) for the installation of this sample system was $3 500 000. This estimate included the cost of the pumps, transmission piping and associated trenching and bedding, loop and hydrant lateral piping and associated trenching and bedding, isolation valves and pits, controls, and cathodic protection. Not included were the costs of storage tanks, site surveying, soils testing, contingencies for special site conditions, paving or pavement removal, electrical power distribution, fuel receipt facilities, hydrant trucks, or fire protection.

LEAK DETECTION METHODS

The existing leak detection methods that appear applicable to hydrant systems, either directly or that could be modified, were reviewed. Each of these are discussed below. Each method is described and an estimate of its expected performance given. The advantages and disadvantages of each method are listed, and an estimate of the cost associated with that method for the sample hydrant system is presented.

Double-Wall Systems

Method Description--One method of constructing the hydrant system distribution piping is to use double-wall pipe. This construction method provides a "pipe within a pipe." If a leak should develop in the inner pipe, the released product would be contained in the interstitial space between the two pipes. This release could be readily detected by any of a number of leak detection methods, such as vapor monitors or liquid petroleum product sensors. The method is being used in some current new construction. The piping layout is often purposely designed to contain periodic high and low points, so that any released product would flow to a low point where it could be detected and withdrawn. This procedure would also help to localize the point of leakage. Thus, the basic method can readily identify a leak and, with appropriate design, also help to determine the location of the leak.

Detection Limit--A double-walled system can detect the smallest of leaks, given that the line break was limited to the interior wall only and time is allowed for the released product to flow to the detector.

Advantages--This construction technique provides a number of advantages over the usual single wall construction. It provides a second level of protection against a leak. Thus, a release of product does not contaminate the environment; any release is contained and can be suitably disposed. Similarly, the size of the leak is relatively unimportant since the product remains in the interstitial space. Secondly, the interstitial space between the two walls provides a means for simplifying leak detection. This space could be monitored by any of several methods for presence of fuel much easier than monitoring the soil outside of the system.

Disadvantages--The obvious drawback is the cost of installation. This method is most feasible for new installations, and is being used at some locations. However, the cost differential over single-wall construction is substantial. Further, retrofitting an existing system is not possible, short of replacing the old system. Double-walled systems may not withstand catastrophic earthquakes, shifting land and other similar occurrences.

Costs--For the sample system, the estimated cost difference for a full double-wall pipe system versus a standard single-wall system is estimated at $4 250 000 for the pipe installation alone. The estimated additional cost for leak monitoring probes at all isolation valve pits and piping low points is $500 000. The cost for monitoring will vary based upon the number and location of monitoring points and the number and location of control/alarm panels associated with the monitoring probes. Installation of product-sensitive cables is discussed later in this paper.

Pressure Monitoring

Method Description--A commonly used method for monitoring an airport hydrant system is to conduct a static pressure test. This is typically done daily when the system is quiet, e.g. from 2 to 4 a.m. The hydrant system pumps bring the system up to normal system operating pressure, typically about 150 psi (1040 kPa). A series of readings of the pressure over some period of time, or a recording of the pressure versus time is made and reviewed by operating personnel the next day. An unusually rapid decay of pressure indicates a leak in the system. Figure 1 is a pressure time plot recorded on a routine basis at an operating hydrant system. The large swings in pressure reflect fueling operations. Note that there are some quiet periods of 2 to 4 hours in length that could be used for leak detection.

Detection Limit--We did not find any reference where a detection limit or minimum detectable leak rate was reported for this method of leak detection. Operators tend to collect months of records and, at least informally, establish norms for the operation of the particular system. We reviewed six months of records from an operating hydrant system with a capacity of 300 000 gal (1 136 000 l) and determined that an operator should easily be able to identify a pressure drop of 20 psi (138 kPa) per hour beyond what is normal. With proper interpretation of local atmospheric temperature data and its effect on fuel volume, this differential could probably be reduced to 10 psi (69 kPa). For this operating hydrant system, this converts to a leak rate of 0.014% or 0.007%, respectively, of system volume per hour, corresponding to volumes of 45 and 22.5 gal/h (170 and 85 L/h). If actual system temperature data were used and a computerized analysis algorithm developed, it is likely that an improvement by another factor of two may be possible.

Advantages--This basic method is rather inexpensive, and is already in place at many installations. If the data it provides are reviewed daily, a major release should be detected quickly, before extensive environmental damage occurs. The method should be applicable to existing as well as new hydrant systems, provided the isolation and hydrant valves are tight.

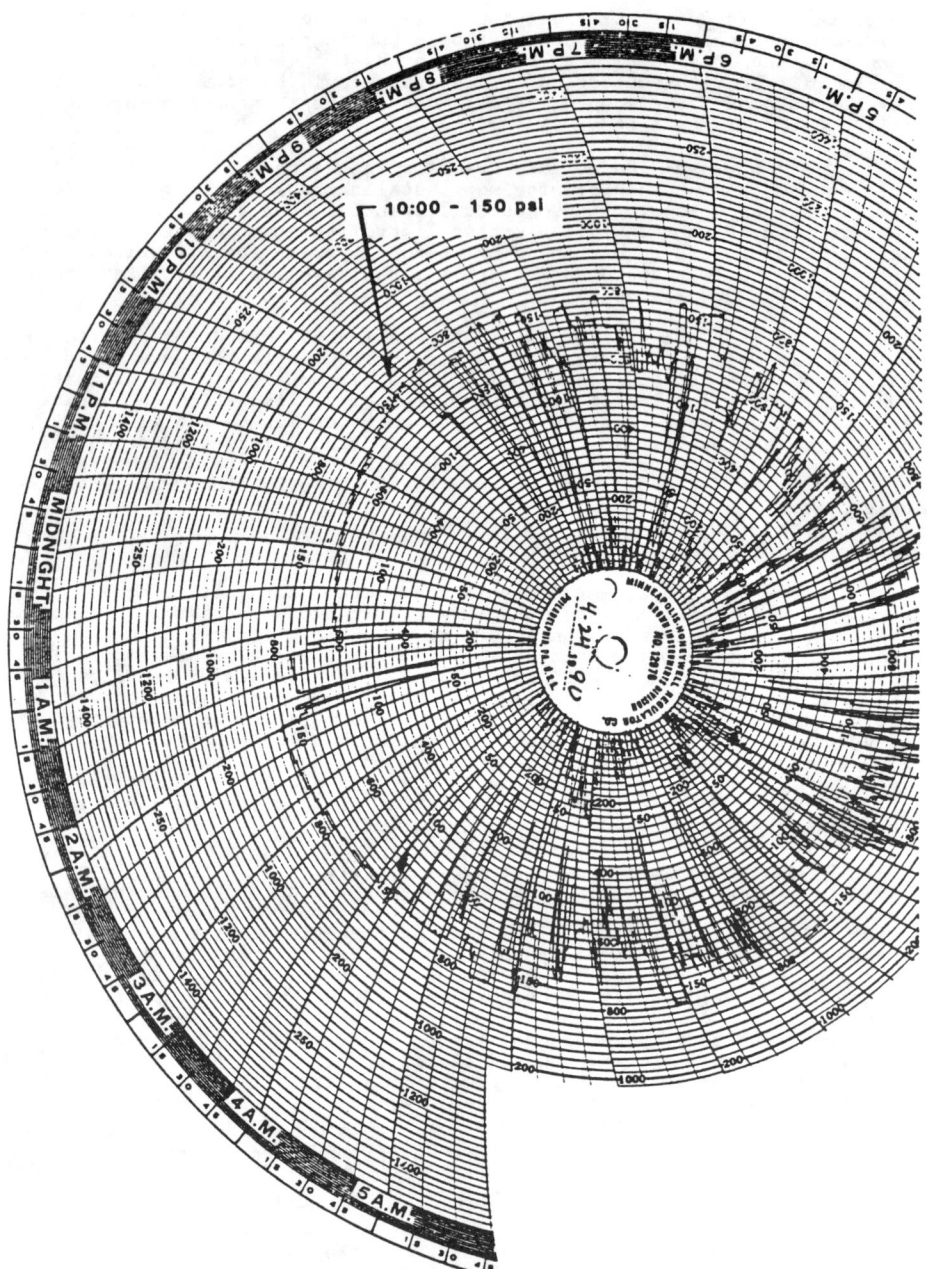

Figure 1 Pressure versus Time Chart

Disadvantages--This method, as presently implemented, cannot detect small leaks. Further, it can only be used on a system in a well maintained condition, with solidly seated valves. To be effective, the

daily records must be reviewed promptly by competent personnel, or
processed by a computerized analysis program. This latter step seems
totally achievable, but is not yet implemented to our knowledge. Also,
pressure tests will only identify the presence of a leak; leak location
techniques must be used in conjunction with this method if a leak is
identified.

Costs--The cost of implementing a pressure monitoring system for
the detection of leaks will vary, depending upon the degree of accuracy
and sophistication required. It can range from the cost of training and
daily review of system records by an individual on a system already
equipped with pressure and flow monitoring devices to $150 000 to
$200 000 for the installation of precision pressure and temperature
monitors and transmitters, data acquisition hardware, and computer
software development and installation to implement a fully automatic
integrated monitoring system. Cost for computer software is estimated
at $50 000 including development and installation.

Precision Pressure Testing

Method Description--There are a number of service station-oriented
methods on the market designed to use pressurization to test piping
systems for leaks. Most of these are designed for the smaller piping
systems associated with underground storage tanks, not hydrant systems.
However, the concepts could also be applied to hydrant systems, but they
would not have the same performance capabilities in terms of minimum
detectable leak rate with hydrant systems as with smaller systems.

The major complicating factor in any pressure test is temperature.
Declining temperatures and leaking systems both lead to decreased
product volume or reduced product pressure. Separating these two
effects is thus a critical element of any pressure test method. The
discrimination between these two effects is essentially this: the
effects of temperature declines tend to taper off as the ambient (soil)
temperature is reached, whereas losses due to a leak remain constant if
the system pressure is constant.

Most of the pressure testing methods for piping systems would also
be applicable to hydrant systems, although those that attempt to
quantify the temperature effect, or that use an algorithm that reduces
or removes the temperature effect, would be most recommended. Also,
precision pressure tests will only identify the presence of a leak; leak
location techniques must be used in conjunction with precision pressure
testing if a leak is identified.

There are number of precision pressure test methods currently
marketed for smaller systems. Because these have not been applied to
hydrant systems, there is no empirical evidence about the detection
limit. We developed a computer model of a pressurized hydrant system to
simulate the operation of various precision pressure testing methods and
provide an estimate of their performance. The model is capable of
calculating the pressure and temperature of the fluid in the hydrant
system as a function of time when the fluid is rejecting heat to the
surrounding earth while it may also be leaking. Several leak detection
methods described below were simulated using this model. All of these
methods rely on a pressure decay curve over time to estimate a leak rate
and to determine whether a leak exists. They differ in how they collect
data and interpret it. All of these methods work reasonably well for
small retail service station pipes, but would need to be adapted to
hydrant systems.

Temperature Compensated Pressure Test--One method marketed for
piping associated with underground storage tank systems uses sensors
that detect both pressure and temperature. The method compensates the

measured pressure for changes in temperature. In a tight system over time the temperature compensated pressure should stabilize to a constant value, indicating that the cause for pressure drops was a decreasing temperature. In a leaking system, the temperature compensated pressure would continue to decline at a nearly constant rate, indicating a leak.

As marketed for small piping systems, the method has only one temperature sensor. For hydrant systems, multiple temperature sensors would be preferable and some algorithm to provide an average temperature would be needed. The time required for a hydrant system to come to temperature equilibrium might be substantially longer than for smaller systems.

<u>Constant Pressure Test</u>--A prominently used piping test method for retail service station piping is a constant pressure test. In this test the piping system is taken out of service and a test apparatus is connected to it. The test apparatus contains a graduated cylinder with product in it. Pressurized nitrogen is used to maintain a constant pressure on the system. A loss of volume in the pipe, whether from a leak or decreasing temperature is made up from the product in the graduated cylinder to maintain constant pressure in the pipe. The amount of product needed to maintain constant pressure is recorded at fixed time intervals.

In a tight system the amount of product used in each time interval to maintain pressure will decrease to zero as the system comes into temperature equilibrium. In a leaking system the amount of product used in each interval will approach a non-zero asymptote, indicating the leak rate at that pressure. As with the previous test approach, the apparatus would have to be modified to accommodate the larger size of the hydrant system.

<u>Product Addition Test</u>--This is a commonly used method for retail service station piping tests. In it, the piping system is filled with product and pressurized to the test pressure, typically 150% of the operating pressure or 40 to 50 psi (375-350 kPa) for these systems. At uniform time intervals of 5 to 15 minutes, the system is brought back to the test pressure and the amount of product added is recorded.

For a tight system the amount of product needed at successive intervals should decline to nearly zero as temperature equilibrium is reached, while with a leaking system the amount would reach a constant non-zero value. The amount per unit time is used to calculate a leak rate that is interpreted as within allowable error or indicative of a leak. Operator judgement is required to determine when temperature stabilization has occurred. The test equipment would have to be modified for use on a hydrant system.

<u>Pressure Cycle Test</u>--This is a method controlled by a microprocessor. The system pressure is raised to the test level, typically 150% of normal operating pressure. The pressure is monitored and the time required to drop to a specified lower pressure is recorded. The system pressure is then raised back to the test level and the cycle repeated. The method uses an algorithm to analyze the successive time intervals needed for the pressure to drop the specified amount. With a tight system the intervals get progressively longer as the temperature changes are reduced. A leaking system is indicated by relatively constant time intervals of fairly short duration.

<u>Time Cycle Test</u>--The time cycle test is similar to the pressure cycle test with the difference being that the pressure is cycled at fixed time intervals. The amount of pressure change in each interval is recorded and analyzed to determine whether a leak is present. A leak would produce nearly constant pressure drops, while a temperature effect

would show progressively smaller pressure drops. Again the pressure decay curve with time is used to determine whether a leak is present and to estimate the size of the leak rate.

Pressure-Step Test--This method uses the most sophisticated pressure curve algorithm. A version of it is marketed by a German firm and is installed in some European airport hydrant systems. Its operation is illustrated in Figure 2.

The pressure of the system is first brought up to a high test level (point A) and then allowed to decline as a result of temperature change. After a fixed period (10 minutes) to allow for system stabilization, the average pressure-loss rate over a two-minute period is determined. Then, at point B the pressure is quickly dropped (stepped) to a lower value such as point C. Again, a ten-minute stabilization period is allowed followed by a two-minute period during which the average pressure-loss rate is determined. Then the pressure is stepped up from point D to point E, its earlier high test level. Once again, a ten-minute stabilization period is followed by a two minute period during which the average pressure-loss rate is determined.

The two average pressure-loss rates after the high-pressure tests (points B and F) are themselves averaged and that value is compared with the average low pressure, pressure-loss rate at point D. If the system is tight (Figure 2a) the two values compared should be essentially equal, assuming the temperature has changed at a nearly linear rate. However, if the system has a leak, as in Figure 2b, the high and low pressure average pressure-loss rates will not be equal because leak rate is a strong function of pressure. Therefore, the leak rate at points B and F will be much larger than at point D because B and F are at a higher pressure.

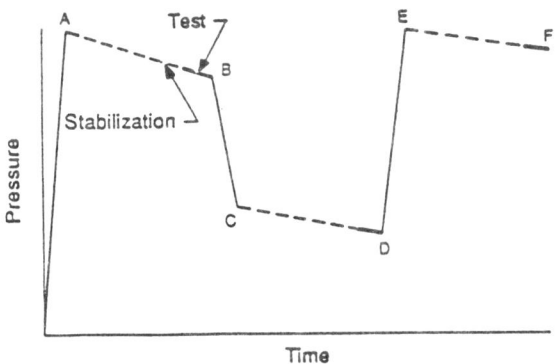

Figure 2a. Pressure-Step Test, Tight System.

Detection Limit--The minimum size of leak that may be detected with one of these precision pressure tests is not known with accuracy. The only one of the above techniques marketed for hydrant system use is the German pressure-step technique. The company guarantees that it can detect a leak of 0.004% of the system capacity, per hour, with a 45-minute test. Further, the inventor claims it can detect 0.002%, so he feels confidant in guaranteeing a 0.004% detection capability.

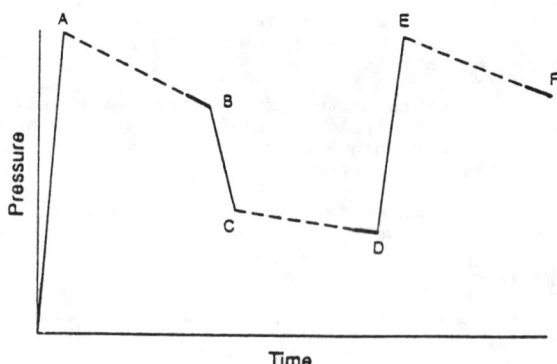

Figure 2b. Pressure-Step Test, Leaking System.

Our computer modeling of the pressure-step technique, including temperature losses, indicates that the 0.004% capability should be easily achievable. However, the computer model does not include "system noise" which would appear as statistical fluctuations in recorded pressure. Adding this noise makes leak detection more difficult, although still achievable with longer tests.

Our review of pressure data such as that shown in Figure 1 indicated that the noise in the data is not excessive, and probably averages less than 1 psi (6.9 kPa), at least after an initial stabilization period. Assuming this to be the case, and extending the test period from 45 minutes to, say, an hour and a half would theoretically enable the pressure-step process to be repeated several times, and three estimations of the leak rate to be obtained. Doing this could bring the detection limit down to about 0.001% of system capacity per hour. For the example hydrant system this would be a leak of about 1.5 gal/h (5.7 L/h) for the entire system. If the transmission line from the storage area were isolated and only the remainder of the system tested, a leak as small as 1/4 gal/h (1 L/h) would be detectable.

Advantages--A precision pressure test system such as the pressure-step method can be made to detect quite small leaks (small for hydrant systems). The technology is available and has been in use at several airports around the world, although not in the U.S. Once installed, it can be applied nightly. The test duration is very short (45 minutes), although we would suggest further experimentation with longer tests, that can still be conducted nightly during the system quiet time. The method can be implemented on an existing hydrant system, although some valving may also have to be replaced. It is even more suited to a new system, where additional isolation valves can readily be installed to separate the system into smaller segments.

Disadvantages--At this time there appears to be only one marketer of a precision pressure test method for hydrant systems. Unless others modify their methods so as to be applicable for hydrant systems, a monopoly exists.

The hydrant system to be tested must be very tight, otherwise false failure results would be obtained. This means that the valves used to isolate different portions of the system must close very tightly, even against the high test pressure. It also means that the hydrant valves must be very carefully maintained, and any leaks repaired immediately.

The presently available pressure-step method is capital intensive. A recent installation at a 20-gate concourse in Europe, similar in size to the example, cost about $1 million. It is expected that this system incorporated some unnecessary features. For example, each of the 40 or so hydrant pits was specially designed and an additional, remotely operated, precision shut-off valve was added to each hydrant. (This was done to assure that weeping at a hydrant would not produce false results.) As mentioned previously, precision pressure tests will only identify the presence of a leak; leak location techniques must be used in conjunction with precision pressure testing if a leak is identified.

Costs--As noted above, a recent system comparable in size to our example system was installed at a cost of $1 million. By keeping the installation to the minimum required, and operating a rigorous maintenance program on the hydrant valves, the initial installed cost could be reduced to about $500 000.

Volumetric Methods

Method Description--The most commonly used means of performing precision tightness tests of underground storage tanks is one of a number of widely marketed volumetric methods. These methods all monitor changes in the amount of product in the tank over a period of time. Some methods measure the level of the product in the tank, the temperature at one or more locations in the tank, and compute the temperature-compensated change in the volume of the product. Other methods employ some variation of a mass measurement concept which is self-temperature compensating. All methods operate at or only slightly above atmospheric pressure. The system is taken out of service while the testing is taking place. The duration of the test period is from 4 to 8 hours, depending on the details of the particular method.

Detection Limit--There is a lack of performance data with respect to accurate volumetric testing, and it is unlikely that this method would be applicable to hydrant systems. It may be possible to take a hydrant system out of service for a few hours, reduce the system pressure to near atmospheric, and monitor the level of product in, say, a standpipe coupled to the system. Moreover, it should be possible to observe level changes in the standpipe equivalent to 0.1 gal/h (0.38 L/h), just as is done with underground storage tanks. The difficulty is compensating for temperature changes.

As noted earlier, an 0.1 gal/h (0.38 L/h) change in volume would be observed with a temperature change of only 0.0013°F, in the 157 000-gal (594 000 l) sample system. It is very difficult to measure temperature changes to this level of accuracy, at a point. However, it seems totally unrealistic to expect to characterize the temperature of the entire 157 000 gal (594 000 l) of product in the system to anywhere near this accuracy.

Metering Methods

Method Description--One way that leaks are detected in high pressure cross country distribution pipelines is comparing flow meter readings at two different points along the pipeline. The method could theoretically be applied to airport hydrant systems by metering the flow of product into the pressurized system, metering all withdrawals from the system because of refueling operations, and converting all measured volumes to net volumes using temperature data. The usefulness of this method will depend on the precision and accuracy maintained.

Detection Limit (Theoretical Application)--Manufacturers of flow meters indicate that the normal sensitivity of calibrated flow meters is about 1%. State regulations for dispensing gasoline or diesel fuel at retail outlets require that the meters be accurate to within 0.5%. This is the same accuracy to which the meters used for airport refueling operations are calibrated. In exceptional circumstances, manufacturers indicate, it might be possible to calibrate a meter to 0.02%. It is unknown at this time as to how frequently such a meter would have to be recalibrated, or how this accuracy would be affected by changes in fuel temperature and flow rate, both of which vary significantly in an operating hydrant system.

If one assumes that with unusual effort one could maintain meter accuracy to 0.1%, what does this suggest for a minimum leak detection level? If a hydrant system were to have such a highly accurate meter at the inlet side (just down stream of the pumps), the flow into the system would be known with a standard deviation of 0.1%. The flow out of the system is measured by meters on a number of hydrant carts. If all of their meters are accurate within a standard deviation of 0.1%, and the errors are normally distributed, the average error in flow measured out of the system is zero. (This is obviously a great simplification, and assumes there is no bias in the collection of meters. This assumption should be studied more closely if this concept is taken seriously.)

Applying the above assumptions to the sample 20-gate concourse leads to the following conclusions. For an average daily throughput of 250 000 gal (946 000 l), the flow measurement would have an error with a standard deviation of 250 gal (9461). A leak, to be reasonably detectable, should be at least two standard deviations, or 500 gal (1890 l) per day (21 gal/h) (79 L/h). This is a leak rate of 0.2% of system throughput (this is also about 0.013% per hour of system volume). Although this is better than the EPA requirement of 1.0% for monitoring methods, it is not as good as what might be expected from a statistical inventory control method, discussed next. Further, the metering method would involve about as much paper work as statistical inventory monitoring, plus significant costs in installing precision meters and keeping them well calibrated. Therefore, this method is not considered further.

Inventory Reconciliation

Method Description--Inventory reconciliation is the process of comparing the physical inventory with an accounting inventory. The two inventory figures are typically referred to as "book" (for the accounting number) and "stick" or "physical" for the measured value. Differences between the two inventory measures are tracked over time to identify discrepancies. The discrepancies may be related to losses from a variety of sources. The potential sources include shortages on delivery, leaks, meter inaccuracies, etc.

The physical inventory is obtained by measuring the physical volume of product in the system. This is done on a daily basis by using a tank gauge on each tank and converting the reading to gal. Any other storage of the product, for example in trucks, is also physically measured. The amount of product in the piping of the hydrant system is generally taken as constant.

The book inventory is the volume that is recorded on the books as on hand. It is calculated by taking the physical inventory for the previous day, adding receipts and subtracting issues. Thus, the book inventory each day is the amount that should be on hand. A comparison of the physically measured inventory with the book inventory provides a means of determining whether any unexplained losses have occurred.

As applied currently, inventory reconciliation is applicable to the entire fuel system, including the storage tanks, hydrant system, and tank trucks used to deliver to or de-fuel planes. In order to do inventory reconciliation, one must be able to measure the amount of product received, measure the amount of product dispensed, and measure the amount of product on hand at any given time. Receipts may be measured by meters in a pipeline or on delivery trucks or by weight of product supplied by truck. Issues are generally measured by flow meter. Inventory on hand is usually measured by a sight gauge or a stick reading of depth of product in a tank or tank truck and converted to volume using a table for the tank.

All of the measurements required are usually made in gross gal (volume at the existing product temperature). A conversion is often made to net gal (volume at a standard temperature of 60°F or 15.6°C). Generally it is desirable to make comparisons in terms of net gal as this removes temperature differences as a source of discrepancy between the physical and book inventory. This requires a measurement of the product temperature or specific gravity so that a correction to net gal can be made.

An example of one month of daily inventory data was provided by a major airline for one of their hub operations. Receipts averaged 600 186 gal (2 272 000 l) per day, issues averaged 600 040 (2 271 000 l) gal per day, and the physical inventory averaged 2 646 534 gal (10 020 000 l) per day. The amount of product needed to fill the hydrant system was given as 433 400 gal (1 641 000 l). All volumes are net. The daily variance between the book and physical inventory was calculated and is plotted in Figure 3. It was usually around 2% of issues. Two days had extremely large daily variances of nearly 50% of issues. These two outliers were opposite in sign, suggesting that a receipt was recorded the wrong day. The two outliers are not shown in the figure. Excluding these outliers, the daily variances had a standard deviation of 10 000 gal (40 000 l) (net) per day. Using a regression model to estimate the trend over the month resulted in a slope of 300 gal (1140 l) per day with a standard error 250 gal (950 l).

Detection Limit--The data analysis described above for one month's data can be used to estimate the leak detection limit using statistical inventory reconciliation methods. If the standard error of about 250 gal (950 l) is representative, then a statistical inventory reconciliation analysis could use about 500 gal (1900 l) per day as a threshold for identifying a leak for this system, and would be able to detect a loss on the order of 1000 gal (3800 l) per day, or 42 gal (160 l) per hour, with about 95% probability, based on analysis of 30 days of inventory data.

The detectable leak rate is often stated as a percentage of daily throughput, and will probably increase with the daily throughput of the system and with the overall system capacity. For the sample hydrant system, the throughput was about 600 000 gal (12 270 000 l) per day, so the method should be capable of detecting 1000/600 000, or 0.167% of throughput. This is 1/6 of the EPA requirement of 1% of throughput for inventory monitoring methods applied to underground storage tanks.

Discussions of these data and our analyses with vendors of statistical inventory services indicate that using more sophisticated analysis methods, and with extra care in the data collection by the operator, an improvement in leak detection capability by a factor of two should be easily attainable. Thus, a leak of 0.083% of throughput should be identifiable. For the sample system described earlier, the 0.083% of throughput translates to 8.65 gal/h, or 0.0055% of system capacity per hour.

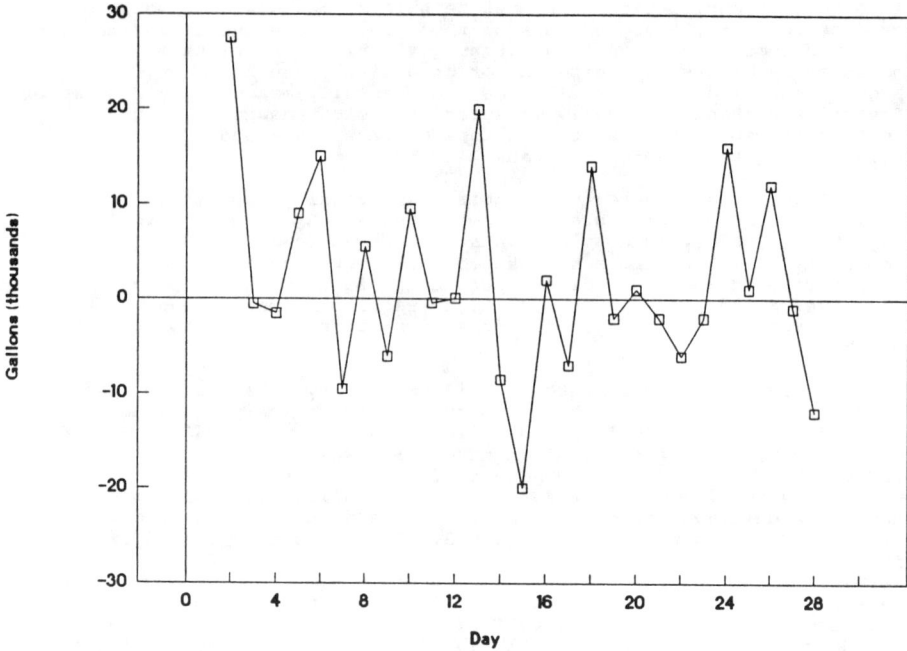

Figure 3 Daily Inventory Variation

<u>Advantages</u>--Operators are already collecting the type of data needed for conducting inventory analyses, but for different reasons. The added costs of refining the effort to improve the quality of the data should be minimal. The method also gives the operator useful information on the financial operations of the fuel hydrant system operation. The method thus requires little change in the way the operator is presently running the system. Thus, it is equally applicable to existing and to new systems. The method also has the advantage that it should be able to detect a leak <u>anywhere</u> in the system, including the storage tanks.

<u>Disadvantages</u>--The method can only detect rather large leaks, even though it can easily meet EPA requirements for inventory control for underground storage tanks. Also, it is only used monthly, and sizeable releases could occur before detection.

The method cannot locate a leak. Since the detected leak could be anywhere in the system, including the storage facilities, locating the leak might be more difficult than with other methods.

Some improvements in data collection by the operator might be required, although they should not be extreme. Meters need to be carefully calibrated and maintained, and level measurement sensitivity in storage tanks may need to be increased. Also, the statistical software might have to be specialized or adapted to each individual hydrant system, at added cost.

<u>Costs</u>--The cost of statistical services to review the inventory data is estimated at $2 000 per month for the sample system. Of primary

importance for this method is the quality of data obtained. The cost of providing this data will vary with the method of operation for each system and each operator.

Tracer Methods

Method Description--Tracer techniques have been used for both soil gas surveys and leak detection. In the latter case, fuel in a tank is spiked with an inert chemical tracer that can be detected in the soil if a leak occurs in either the tanks or the product lines. If spiked fuel is released, the tracer volatilizes and migrates throughout the area around the release. Sampling is usually conducted two or three weeks after spiking to allow tracer to diffuse to a sample probe in the case of a leak. Vapor samples are collected at various points along the piping and analyzed for the presence of tracer using gas chromatography. The spacing of the sample probes is site-specific, but is typically 15 to 25 feet (4.6 to 7.6 m). Since the source of the tracer is unique, the presence of tracer in the environment around the line would be considered as conclusive evidence of a leak. Absence of tracer would be considered as evidence that the line was tight. Spills or other releases of the fuel containing the tracer would also be detected.

Tracers have proven effective in detecting even small leaks (measured in parts per trillion in backfill), and are not affected by fluctuations in product temperature or other factors that complicate volumetric leak detection methods. Detection of release of a few liters is readily achievable under normal test conditions.

In principle, any volatile compound which is not normally found in the product of interest can be used as a tracer. For most current applications halocarbons are used. Although some concern has been expressed about their use in aircraft engines, Air Force testing has lead them to conclude that there are no detrimental effects from the low concentrations required for tracer testing.

The leak detection method is not a continuous one, however, because sampling for tracer in the soil is performed periodically, probably annually for an airport hydrant system. Sampling probes may be left in the ground for reuse. For large installations that require more frequent monitoring, groups of probes may be manifolded together in an aspirated system. Pipeline systems have also been tested using a porous leak detection hose, up to 400 ft (122m) in length, that is installed in the soil above the pipeline. The hose replaces the probes for purposes of soil gas collection. Retrofitting of the hose for existing hydrant systems could be difficult, especially under concrete paving.

Detection Limit--It is expected that releases smaller than 0.1 gal/h (0.38 L/h) can be detected using the tracer method if probes are properly spaced.

Advantages--Tracer methods have been successfully implemented for leak detection purposes in military hydrant systems. The hydrant system can be tested while in use, which eliminates down time. The presence of tracer in the soil gas unambiguously defines a leak, because the halocarbons are not found in natural soils. The method can also pinpoint leak locations accurately. Leaks may be distinguished from previous spills, and the method may also be used in areas where soils have high product vapor background levels.

Disadvantages--The most serious limitation for the use of tracers is the lack of aircraft manufacturers' approval for adding the tracer to the fuel. In addition, neither ASTM or API have approved the addition of tracers to aircraft fuel. Another potential disadvantage is that a high water table may interfere with the detection of the tracer. Also, after one tracer has been released into the backfill, another tracer must be used or there is a problem of background level of the tracer to contend with. Also, it is only used annually, or monthly at best, and sizeable releases could occur before detection.

Costs--The costs to use tracer as a leak detection method are highest for the initial application of the method, including the design and installation of sampling probes. Subsequent sampling can then use the same sample probes at a greatly reduced cost. The installation of sampling probes is approximately $25 each with a sample analysis cost of $25 each. Lines which are not under concrete or asphalt can be retrofitted with a long porous tube for sampling purposes. The total cost for installation, sampling and analysis for the sample hydrant system is estimated to be about $35 000 for installation of the system and about $11 000 for annual sampling and analysis.

Vapor Monitoring

Method Description--Vapor monitoring is based on the detection of petroleum soil gas vapors. Two somewhat different approaches have been developed. Passive systems locate a sensor that responds to hydrocarbon vapors in each soil gas monitoring well. These sensors do not typically provide concentration levels, but function as pass/fail devices which alarm at a preset level. In contrast, aspirated systems pull samples from the well to a central sensor. A valving system is used to sample several wells sequentially at set intervals. These types of devices do report concentration levels.

The product must be volatile enough to yield vapor concentrations within the range of the sensor response and the surrounding soil must be porous enough to permit migration of the vapors to the well. Typically, vapor monitors are quite sensitive to releases, but have a tendency to produce false alarms (declaring a leak when none is present) if the data are not properly interpreted. Since there is some natural variability in the background levels of hydrocarbons, the identification of a leak requires that trends in the vapor concentration be recognized. For example, rain may cause a temporary rise in hydrocarbon vapor levels which will trigger an alarm. Again, background concentrations can cause problems for these types of sensors although most have provisions for off-setting these effects.

An aspirated system developed by one company is discussed to illustrate how a vapor monitoring system might be adapted to hydrant systems. A series of monitoring wells is installed near the product lines. Vapor samples are drawn from the wells through transport tubing to an indoor console for analysis by the single sensor. After three consecutive readings exceed a set threshold, an alarm is triggered. As vapors migrate through the backfill, they are detected by other monitoring wells, and concentrations of product in parts per million are determined at each well.

An alternative approach has been developed which utilizes a permeable tube which is installed alongside the piping. Hydrocarbons migrate into the tube and are aspirated at periodic intervals into a detector located at one end of the tube. An approximate location of the leak can be obtained from the time interval necessary for the hydrocarbon to reach the detector. The method has been successfully applied to pipelines and could be adapted to new hydrant system installations.

Detection Limit--Continuous vapor monitors are capable of detecting releases of only a few gal if wells are properly spaced.

Advantages--Vapor monitoring has been one of the more sensitive methods for detecting releases. The method is capable of detecting releases quickly, particularly if the soil is permeable. Some low cost vapor monitoring methods are available which utilize absorption tubes or portable vapor detectors.

Disadvantages--The method can only be applied where the background vapor levels are maintained at low to moderate levels and the soil is permeable. The product must be volatile enough to vaporize, and the detector must be capable of responding to the species of interest. A high water table may interfere with the detection of the vapor. The interpretation of the vapor sensor data requires a fairly high level of expertise.

Installation of the numerous wells is costly, and there may be difficulties detecting small leaks in areas with high background vapor concentrations. Retrofitting older systems would be difficult due to the requirement for installation of electrical cables (for passive systems) or sample lines (for aspirated systems). Another limitation of aspirated systems is the need for aspirated lines that may be several hundred feet long. It could take a long time for the vapor to travel to the sensor, and diffusion through the tubing, or adsorption/desorption in the tubing, may also take place along the length of the line. The installation of horizontal monitoring wells could also be difficult to achieve in an airport setting.

Costs--There are currently no installations of this equipment at similar facilities, and it is difficult to see how such a monitoring system could be economically retrofitted to an existing hydrant system. The installation costs would likely be many times the cost of the instrumentation. Installation costs for a horizontal well are unknown, but would likely be high for existing lines under concrete. A series of monitoring wells at 50-foot (15m) intervals could cost as much as $100 000 to $200 000 for the example airport. Monitoring equipment could cost an additional $130 000 to $150 000. Permeable tubing costs are approximately $4 per foot ($13 per m) plus installation and monitoring equipment costs. For the example system, the cost of tubing alone would be $56 000.

Groundwater Monitoring

Method Description--Groundwater monitoring is based on the detection of free product floating on the surface of the water table. In the simplest case, the condition of the groundwater can be visually inspected for petroleum with the aid of a clear, acrylic well bailer. Some methods use conductivity meters and product-soluble devices to detect the presence of petroleum. EPA requires that as little as 1/8 in. (0.32 cm) be detectable, which is close to the detectable limit using manual "sticking". The more sophisticated systems utilize automated equipment which have the capability of detecting smaller amounts of petroleum.

This is a qualitative method since the volume of petroleum which enters the well is determined by a number of factors including soil permeability, hydrogeologic gradient, depth of groundwater and the differential in the hydraulic pressures. In addition, the volume of petroleum required to trigger an alarm varies with method design. Thus, alarm indications of both the visual inspection and the automated detection systems may represent a considerable loss of product over a long period of time.

Detection Limit--The method is not quantitative and the volume of release required to trigger an alarm varies widely with soil type, groundwater level, product type and type of monitoring system. A visual inspection will detect the presence of a sheen on the surface. The development of a 1/8 in. (0.32cm) layer of product could represent a considerable loss of product over a long period of time.

<u>Advantages</u>--Groundwater monitoring has been widely used in some areas of the country, particularly where high groundwater is present. Under conditions where the groundwater is within a few feet of the surface and the soil is permeable, this method may be appropriate. If the monitoring well is checked by bailing on a regular basis, it can be a low cost monitoring method as well.

<u>Disadvantages</u>--The current EPA underground storage tank regulations permit groundwater monitoring only if the water table is not more than 20 ft (6.1m) below grade. When product is detected on the groundwater, a release has occurred and may require remediation. It is also difficult to locate the leak.

One of the most serious difficulties in using groundwater monitoring for hydrant systems involves the number of wells which would need to be installed to provide a reasonable expectation for detecting a leak in a timely manner. The cost to install a monitoring well is in the range of $1 500 to $2 000 each for a normal installation.

It is likely that wells would need to be located at intervals of around 25 to 50 ft (4.6 to 7.6m) (less in some environments). This could be cost prohibitive in some instances where long hydrant line segments are involved. Even with many wells, considerable product might be released before there was sufficient migration to a well to trigger an alarm. In addition, the presence of features such as fractured rock, clay lenses, or the presence of underground utilities may route the product away from the monitoring well so that it is never detected.

Other factors to be considered include the fact that alarms may also be triggered by releases not related to a line leak, so that additional confirmation tests would need to be conducted before a leak could be reliably declared. Automated systems also require periodic maintenance to verify proper operation. Fouling or corrosion of the sensor may render the system inoperative.

Finally, once a well has been contaminated with product, its usefulness as a monitoring well is seriously compromised. Remediation of the site would then be necessary to restore the utility of the well for monitoring purposes.

<u>Costs</u>--The installation of a ground water monitoring well can run as high as $1 500 to $2 000 each. To install monitors at 50-foot (7.6m) intervals would require the installation of 260 monitoring wells for the example airport. Costs for installation are estimated to range from $350 000 to $500 000, not including the monitoring equipment.

<u>Product Sensitive Cables</u>

<u>Method Description</u>--Product-sensitive cables are constructed of materials which will degrade or change electrical properties when they come in contact with hydrocarbons. This change can, by digital read-out instrumentation, be traced to the approximate position of the failure. The surrounding environment must also be cleaned up completely to prevent the destruction of the new cable. Various cables have been designed to detect fuels, solvents, and aqueous chemicals. The cables are composed of a signal wire, a continuity monitoring wire, and semiconductive jacketed sensors enclosed within a fluoropolymer braid. Cable is installed in a double-contained system near the product lines or tanks. In some cases, petroleum destroys the usefulness of the

monitoring cable, which must then be replaced in the area where the leak occurred.

One company is currently developing a product that can be directly buried, and has expressed an interest in adapting it to airport hydrant systems. Reportedly, these product-sensitive cables can be installed in either the interstice of a double-walled hydrant line or in a slotted retainer pipe buried beneath the hydrant line.

Detection Limits--When used in a double-walled pipeline, or in a separate slotted retainer pipe buried beneath the hydrant line, very small releases (probably less than one gal) would be detected quickly.

Advantages--Product sensitive cable can be an extremely sensitive form of release monitoring, particularly when located in the annular space of a double walled pipe or in a lined trench. They are best applied to new installations where they can be designed into the system and installed along with the piping. In these cases, the background levels of product will be low so that there will be minimum interferences.

One of the most attractive features of product-sensitive cables is that they can pinpoint the location of the leak. This could be of considerable advantage for remediation where lines are located under concrete or buildings.

Disadvantages--Retrofitting product cables to an existing system may not be technically or economically feasible. First, the installation costs would be substantial. Each line would require a trench to be excavated very close to it. The cable would then be laid below the piping (directly under in the 6 o'clock position if possible), the trench filled in, and the surface restored. This could be very disruptive to operations during installation and would be difficult to achieve under thick concrete flight aprons.

A second major problem is that existing contamination and future surface spillage could render the system useless unless it is economically feasible to replace the segment of damaged cable and thus recondition the monitoring system. Past spills and leaks would need to be completely remediated prior to installation of the cable to prevent alarms from occurring and the necessity for replacement of the cable.

Costs--The cost of this type of cable is estimated to be $10 per foot. For the example airport, cost of the cable would be approximately $130 000 plus installation. Retrofitting an existing hydrant system with cable would involve excavation along the entire length of the hydrant system, probably a cost-prohibitive venture.

Acoustical Methods

Method Description--Acoustical methods are based on identifying the sounds produced when a fluid flows through a hole under pressure. Microphones placed on the pipe surface or in the liquid within the pipe are used to monitor this noise. Electronic devices are then used to filter unwanted sounds from the noise to identify the leak. The use of multiple microphones allows the approximate position of the hole to be determined.

Acoustical methods have been used primarily on steam lines and water lines. Further development will be necessary to determine the probability of success for lines containing hydrocarbons.

The permanently installed systems show the most promise for applicability to airport hydrant systems. Additional research must be conducted to determine what size of a leak can be detected on pipes of various diameters and compositions.

Detection Limits--The sensitivity of acoustical methods is limited. Leak rates of one gal/h (3.8 L/h) seem to be at the lower limit of detection. The distance from which the leak can be detected is unknown, but it seems that at the pressures under which a typical hydrant operates this distance may be short. According to one vendor, the prospects for increasing sensitivity are poor.

Advantages--The primary advantage of the acoustical method is in its use in locating the position of a known leak. Acoustic methods of leak detection have been successfully applied to water and steam pipes, particularly on large diameter piping. Attenuation methods, in which leaks are located by analyzing the signal ratios received from two sensors, may be used.

Poor signal to noise ratios can be accommodated using the cross-correlation acoustical method. In this method, signals from one sensor are compared to signals from another sensor, using an averaging technique. Signal time delay is converted to a leak location. Significant improvements have been made in the technique with the availability of more powerful and faster computers.

Disadvantages--There are a number of possible interferences which can make it difficult to identify a leak. Other sources of noise such as pumps, traffic, or valves in the line may produce sounds when liquid is flowing. There are, however, both electronic hardware and computer programming which can be used to remove these sources of error. Hardware and software are large investments in this method, and the frequency of operation is limited by the speed of the digital processing hardware.

Small-diameter buried pipe and pipes made of cast iron or composite materials have been found to cause high attenuation rates at any given frequency. Multiple leaks between sensor pairs cannot be located using acoustic methods.

Costs--Since this method has not been tested or applied to hydrant fueling or similar systems, no costs have been assigned to the sample airport. Current hardware which has been used on water lines is available for $20 000 to $55 000, depending on the sophistication of the system. Costs to do preliminary testing for feasibility purposes would range from $5 000 to $10 000. Adaptation of the method for continuous monitoring would be costly and would involve a major research and development effort.

Emerging Technology

There is a continual evolution of new ideas for leak detection. Several promising developments are on the horizon which could potentially be applied to hydrant systems. One of these is the development of fiber-optic sensors which can be tuned to respond to specific types of hydrocarbons (e.g., gasoline, Jet A, JP-4, diesel, etc.). This technology may be simpler and therefore lower in cost than some of the existing methods for sensing hydrocarbons.

Better computer software to improve the automatic identification of releases is also under development. So called intelligent probes may make many of the current methods more reliable and user friendly.

A number of manufacturers are considering the use of remote data links between the sensor and console. This would eliminate the need for hard wiring, making the installation of the sensors much simpler. The concept of a remote data link can be easily applied to many of the existing technologies. While additional costs for the electronics are likely, the savings on installation costs should make the total package attractive. This technology is still in the development stages, but it is definitely worth watching.

Summary

Much of the information on leak detection methods for underground airport hydrant systems is summarized in Table 1. This table provides a means of comparing the costs, advantages, and disadvantages of the various methods.

The methods felt to not be applicable to hydrant systems, as discussed in the previous sections, are not included in the table. These include volumetric, metering, and acoustic methods as well as product-sensitive cables (other than their use as a detector in an interstitial monitoring system).

The detection limits are based on information from manufacturers plus engineering calculations and judgement. Many of them are given as an hourly leak rate expressed as a percentage of the system volume. The sample hydrant system has a volume of 157 000 gal (973 000 l). For this system, the pressure monitoring method, for example, should be able to detect a leak rate of 0.01% of volume or about 16 gal (60.6l) per hour. Several of the methods detect leaks by using sampling locations that are closely spaced. The cost estimates for these are based on appropriately spaced sampling locations.

The cost figures are, by and large, installation plus out-of-pocket costs. They do not include the costs of staff time to operate or monitor the systems or to analyze the data.

TABLE 1--Comparison of Leak Detection Options

Option	Detection Limit[1]	Advantage	Disadvantage	Cost ($000)[2]
Double Wall Construction	([3])	a,b,c,d	A,B	5 000
Pressure Monitoring	0.01	e,f,g	C,D,E	[4]250
Precision Pressure Testing[5]	0.002	d,e,h	C,F,G	500
Inventory Reconciliation	0.006[6]	d,f,g,i	D,H,I,J	[7]25
Tracer Methods	([3])	a,c,d,f,j	G,H,K,L	[8]35
Vapor Monitoring	([3])	a,c,d,e	B,L,M,N	[9]300
Ground Water Monitoring	([10])	c,d,e	D,F,O	500

[1]Percent of system capacity per hour.
[2]For sample 20-gate concourse (1990 U.S. dollars)
[3]Not readily quantifiable and not dependent on system capacity. Trace amounts can be detected.
[4]$150 000 to $200 000 for initial installation, if not already in place, plus $50 000 for software and improvements.
[5]Example given is the pressure-step method.
[6]Percent of throughput (0.083).
[7]Doesn't include costs of improved data collection equipment and personnel.
[8]Installation cost only. Add $11 000 per year for annual sampling.
[9]For new installation. Retrofit would be much more expensive, if even feasible.
[10]Not readily quantifiable and not dependent on system capacity. Can detect 1/8" (0.32cm) of product floating on water table.

Advantages

a. Can detect vary small leaks, given enough time.
b. Prevents loss of product to the environment.
c. Can aid in leak location.
d. Existing or proven technology.
e. Can be used daily, to rapidly detect large leaks.
f. Relatively inexpensive.
g. Already in place for many systems, or can be easily added.
h. More sensitive than other direct measuring methods.
i. Can detect losses from entire system, including ASTs, refueler trucks, etc.
j. Distinguishes new leaks from contamination due to old spills, etc.

Disadvantages

A. Very costly.
B. Retrofitting to existing system difficult or not feasible.
C. Isolation and hydrant valves must be tight.
D. Not as sensitive as some other methods.
E. If not computerized, requires knowledgeable personnel to operate.
F. Quite costly.
G. Only one vendor currently available.
H. Can only be applied infrequently (monthly or annually).
I. Requires accurate record keeping.
J. Less capable than other methods of locating a detected leak.
K. Tracer materials not accepted in jet fuels by commercial airline industry.
L. May not be applicable in presence of high water table.
M. Detection difficult in presence of high background vapor concentration.
N. Interpretation of data requires expert judgement.
O. Only applicable if water table is 20 ft (6.1m) or less from surface.

REFERENCES

United States Federal Register (1988) **53**, 37196-37212.

"Guide Specifications for Airport Hydrant Systems," Air Transport Association of America (1990) A1-A31.

Eric G. Eckert[1], Joseph W. Maresca, Jr.[2], Robert W. Hillger[3], and James J. Yezzi[4]

LOCATION OF LEAKS IN PRESSURIZED PETROLEUM PIPELINES BY MEANS OF PASSIVE-ACOUSTIC SENSING METHODS

REFERENCE: Eckert, E. G., Maresca, J. W., Jr., Hillger, R. W., and ؟ J., Yezzi, J. J., "Location of Leaks in Pressurized Petroleum Pipelines by Means of Passive-Acoustic Sensing Methods," Leak Detection for Underground Storage Tanks, ASTM STP 1161, Philip B. Durgin and Thomas M. Young, Eds., American Society for Testing and Materials, Philadelphia, 1993.

Abstract: Experiments were conducted on the underground pipeline at the EPA's UST Test Apparatus in which three acoustic sensors separated by a maximum distance of 38 m (125 ft) were used to monitor signals produced by 11.4-, 5.7-, and 3.8-L/h (3.0-, 1.5-, and 1.0-gal/h) leaks in the wall of a 5-cm-diameter pressurized petroleum pipeline. The range of line pressures and hole diameters used in the experiments were 70 to 140 kPa (10 to 20 psi), and 0.4 to 0.7 mm (0.015 to 0.030 in.), respectively. Application of a leak location algorithm based upon the technique of coherence function analysis resulted in mean differences of approximately 10 cm between predicted and actual leak locations. Standard deviations of the location estimates were approximately 30 cm.

Spectra computed from leak-on and leak-off time series indicate that the majority of acoustic energy received in the far field of the leak is concentrated in a frequency band from 1 to 4 kHz. The strength of the signal within this band was found to be proportional to the leak flow rate and line pressure. Energy propagation from leak to sensor was observed via three types of wave motion: longitudinal waves in the product, and longitudinal and transverse waves in the steel. The similarity between the measured wave speed and the nominal speed of sound in gasoline suggests that longitudinal waves in the product dominate the spectrum of received acoustic energy. The effects of multiple-mode wave propagation and the reflection of acoustic signals within the pipeline were observed as non-random fluctuations in the measured phase difference between sensor pairs.

Keywords: leak location, leak detection, acoustics, pipelines, underground storage tanks, passive-acoustics, acoustic emissions

---

INTRODUCTION

Millions of underground storage tanks (USTs) are used to store petroleum and other chemicals. The underground pressurized pipelines associated with USTs containing

---
[1] Research engineer, Vista Research, Inc., 100 View Street, Mountain View, CA 94041.
[2] Staff scientist, Vista Research, Inc., 100 View Street, Mountain View, CA 94041.
[3] Environmental scientist, U.S. Environmental Protection Agency, Releases Control Branch, Risk Reduction Engineering Laboratory, Edison, NJ 08837
[4] Senior environmental engineer, U.S. Environmental Protection Agency, Releases Control Branch, Risk Reduction Engineering Laboratory, Edison, NJ 08837

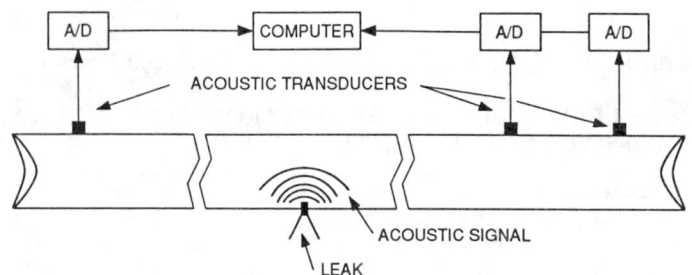

FIG. 1 – Example of a passive-acoustic leak location system.

petroleum motor fuels are typically 5 cm (2 in.) in diameter and 15- to 60-m (50- to 200-ft) in length. These pipelines typically operate at pressures of 140 to 210 kPa (20 to 30 psi). Longer lines, with diameters up to 10 cm (4 in.), are found in some high-volume facilities. There are many systems that can be used to detect leaks in underground pressurized pipelines. When a leak is detected, the first step in the remediation process is to find its location. Passive-acoustic measurements, combined with advanced signal-processing techniques, provide a nondestructive method of leak location that is accurate, relatively simple to perform, and can be applied to a wide variety of pipelines and pipeline products. The concept of using passive acoustics to determine the spatial location of leaks has been around for some time, but this approach has not been applied to underground pressurized petroleum pipelines.

While it is known that a pressurized underground pipeline that is leaking emits an acoustic signal, the strength and characteristics of the signal associated with the leak are not well known. Acoustic systems have been successfully used to detect and locate leaks in nuclear reactors for many years [1]. By means of a cross-correlation analysis, 100- to 400-kHz acoustic sensors spaced at 5- to 10-m intervals can be used to detect leaks of approximately 230 L/h (60 gal/h) with an accuracy that is within 0.5 m. A similar approach has been tested for locating water leaks in 10- to 25-cm (4- to 10-in.)-diameter underground district heating and cooling pipes [2]. Theoretical predictions based on [2] suggest that leaks of 450 L/h (120 gal/h) could be pinpointed to within several meters with sensors spaced at several hundred meters. Using monitoring frequencies less than 25 kHz makes this wider spacing possible; frequencies between 1 and 5 kHz appear to give the best results. Interestingly, leaks that occurred in a steel pipe covered with insulation material (urethane and a rubber jacket) showed a higher level of signal intensity than leaks that occurred in an uncovered pipe.

Figure 1 shows a simple representation of a passive-acoustic leak location system in which three transducers simultaneously sample the acoustic signal. The output of each transducer is digitized and stored as a time series. These time series, recorded by spatially separated sensors, then serve as input to a leak location algorithm. The primary function of the location algorithm is to estimate the time delay between acoustic leak signals received by pairs of sensors. The measured time delay can be used to estimate the source location (for signals received by sensors bracketing the leak) or the propagation speed of the acoustic waves (for signals received by non-bracketing sensor pairs).

Location algorithms that measure the time delays by means of cross-correlation analysis work well provided that the signal is very strong or that the background noise is not excessive. When the acoustic signal is weak in relation to the level of background noise or has a finite frequency bandwidth, more sophisticated signal processing techniques are available. One such technique is *coherence function analysis*.

If the correspondence between received signals is frequency-dependent, or if the phase dependence of the correspondence is a nonlinear function of frequency, the application of coherence function analysis is the means by which the source of the signal is best located. For the purpose of signal estimation and source location, coherence function analysis represents a significant improvement over correlation analysis [3]. Advanced signal processing is required for the successful application of this technology to the problem of leak location for UST pipelines. This paper presents the results of leak location estimates obtained through application of a location algorithm based upon coherence function analysis, and a brief summary of the physics associated with pipeline leak location. A more detailed presentation of these results can be found in [4].

LOCATION OF A CONTINUOUS LEAK SIGNAL

Two criteria must be satisfied in order that accurate location estimates result from the application of the location algorithm: (1) the received signals must originate primarily at a single, localized source and propagate as plane waves along (or within) the pipeline, and (2) the received signals must maintain a reasonable degree of similarity over the maximum sensor separation. If criterion (1) is satisfied, the difference in phase between received waves of a given frequency is simply related to the time delay between signals that arrive at the different sensor locations. The accuracy with which the time delays can be measured is related to criterion (2). The similarity between signals emitted from a localized source and received at separate locations is determined by the signal strength relative to ambient noise (i.e., the signal-to-noise ratio) and the difference in propagation path between the source and each sensor. Due to the complex manner in which the acoustic leak signal is produced (turbulent flow, cavitation) and the many variations in the propagation medium (valves, branches, reflective ends), the degree of signal similarity is not uniform over a broad range of frequencies. Though the signal-to-noise ratio (SNR) provides a reasonable estimate of the frequency band for which accurate leak locations may be obtained, a more sensitive measure of signal similarity is required for the location of small (e.g., 10 L/h or less) leaks.

Consider two time series of acoustic signals, $m_1(t)$, and $m_2(t)$, where each represents the sum of a desired acoustic leak signal, $s(t)$, and a contaminating noise component, $n(t)$. The contaminating noise component could be a combination of ambient acoustic noise in the measurement environment that is uncorrelated at the separated sensors, and electronic noise associated with the data acquisition system. The coherence function, $\gamma^2(f)$, is the normalized cross spectrum of the two measurements,

$$\gamma^2(f) = \frac{\overline{M_1(f)M_2^*(f)}}{\sqrt{\overline{|M_1(f)|^2}}\sqrt{\overline{|M_2(f)|^2}}}, \qquad (1)$$

where the upper-case letters denote the Fourier transform of the respective quantities and the overbar denotes the ensemble average. The magnitude of the complex coherence function measures the similarity between signals $m_1(t)$ and $m_2(t)$ received at spatially separated sensor locations. The coherence phase, $\phi(f)$, measures the relative time delay between the two signals as a function of frequency. The coherence function ranges in magnitude from 0 (signals completely uncorrelated) to 1 (signals completely correlated). Values of $\gamma^2(f)$ exceeding 95% of the noise fluctuations are usually taken as indicating a reliable phase measurement.

If the acoustic leak signal is approximated as a collection of propagating acoustic plane waves that obey the simple linear dispersion relation

$$2\pi f = kV, \qquad (2)$$

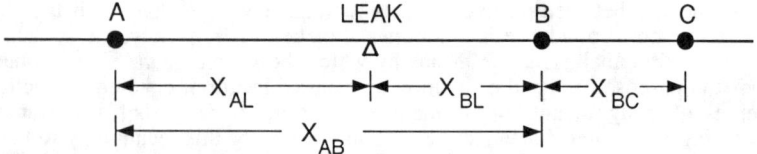

FIG. 2 – Three-sensor approach to acoustic location of leaks.

where k is the wavenumber and V is the propagation speed, the differential separation between two sensors, Δx, and the frequency-dependent phase, $\phi(f)$, are simply related by

$$\phi(f) = 2\pi f\left(\frac{\Delta x}{V}\right). \tag{3}$$

Through the use of coherence function analysis, it is possible to isolate portions of the acoustic spectrum within which the linear dispersion relation is obeyed. The measured phase shift, $\phi(f)$, within these frequency bands can then be used to estimate either the propagation speed of acoustic waves or the differential sensor separation. Because the coherence phase is confined to the range $-180° \leq \phi \leq 180°$, the measured phase generally differs from the actual phase by an unknown factor of 360°, except at very low frequencies and/or very small sensor separations. As a consequence, the measured phase cannot be accurately unwrapped except within frequency bands where $\gamma^2(f)$ is high; thus, a differential form of Eq. (3) must be used to relate sensor separation, propagation speed, and coherence phase:

$$\frac{d\phi}{df} = \frac{2\pi \Delta x}{V}, \tag{4}$$

in which it is assumed that the medium is nondispersive.

The three-sensor approach illustrated in Figure 2 is used to locate leaks in an underground pipeline. Sensor pair B-C is used to measure the *in situ* wave speed, while sensor pairs A-B or A-C are used to estimate the leak location. Because the wave speed associated with a particular product and pipeline geometry is usually unknown, an experimental estimate of the wave speed improves the accuracy of the leak location estimate.

Application of Eq. (4) to sensor pair A-B, which bracket the leak, yields a simple relationship between measured phase, wave speed, and leak location:

$$X_{AL} = \frac{X_{AB}}{2} - \frac{V}{4\pi}\frac{d\phi_{AB}}{df} \tag{5}$$

$$X_{BL} = \frac{X_{AB}}{2} + \frac{V}{4\pi}\frac{d\phi_{AB}}{df} \tag{6}$$

where the subscript L denotes the location of the leak. The wave speed is estimated from the measured phase between sensor pair B-C:

$$V = 2\pi X_{BC}\left(\frac{d\phi_{BC}}{df}\right)^{-1}. \tag{7}$$

The one-standard-deviation uncertainty in the location estimate, $\sigma(X_{AL})$, associated with an ensemble of measurements $\{X_{AL}\}$ obtained through application of Eqs.(5) and

(7) is related to the uncertainty in the derivative of the measured coherence phase and the sensor geometry by:

$$\sigma(X_{AL}) = \frac{\overline{V}}{4\pi}\sigma_s(1 + K^2)^{1/2}, \qquad (8)$$

where

$$K = \frac{(X_{AB} - 2\overline{X}_{AL})}{X_{BC}}. \qquad (9)$$

σ_s is the uncertainty associated with the measurement of the phase-derivative, $d\phi/df$; $\overline{V}$ and $\overline{X}_{AL}$ represent ensemble average values of the propagation speed and leak location, respectively. Two important observations should be made regarding Eq. (8): (1) errors in the measurement of $d\phi/df$ translate directly into errors in location estimate, and (2) the magnitude of the predicted location error is affected by both the overall sensor geometry and by the position of the leak relative to the bracketing sensor-pair. For a given uncertainty in the phase-derivative, σ_s, the location error is minimized when the leak is positioned midway between the bracketing sensors. If $d\phi/df$ is calculated by applying a linear regression to n data points, $\{\phi_i, f_i\}$, contained within a frequency band Δf, σ_s is given by [5]:

$$\sigma_s^2 = \frac{N[1 - \gamma^2]}{2n\gamma^2[N\sum f^2 - (\sum f)^2]} \qquad (10)$$

where N is the number of independent data segments used to compute the coherence function and γ^2 is the amplitude of the coherence function within the analysis frequency band. The predicted error in location estimate obtained by combining Eqs. (8) and (10) can be evaluated for the experimental sensor/leak geometry and the coherence parameters (γ^2 and N) used in the data analysis. Setting $\overline{V}$=1000 m/s, X_{AB}=30 m, X_{BC}=8 m, $\overline{X}_{AL}$=15.5 m, N=15, Δf=200 Hz, n=20, and γ^2=0.35 (95% level of statistical significance for N=15 segments), the predicted one-standard-deviation in location estimate is 8.3 cm. A similar calculation in which sensor pair A-C (X_{AC}=38 m) is used to estimate the leak location yields a location error of 11.0 cm. A detailed discussion of the errors inherent in the location algorithm is provided in [4].

EXPERIMENT DESIGN

The experiments were conducted on the pressurized 5-cm (2-in.)-diameter steel pipeline at EPA's UST Test Apparatus in Edison, New Jersey. A diagram of the UST Test Apparatus pipeline is shown in Figure 3. Access ports required for the attachment of transducers to the pipeline were located at intervals of approximately 8 m (25 ft). The sensor positions shown in Figure 3 were used during all experiments reported in this paper. The transducers chosen for this work were CTI-30 resonant sensors. Though the CTI-30 is designed primarily for acoustic emissions applications, its sensitivity at low frequencies (1-5 kHz) is adequate for the detection of acoustic leak signals in pipelines. The acoustic signals were amplified by 80 dB, in two stages, using battery-operated Panametrics 5660-C preamplifiers and line-driven Krohn-Hite 3342 amplifying filters. A Western Graphtec TDA-3500 transient recorder was used to digitize the acoustic waveforms at a sampling rate of 10 kHz. Data were stored and analyzed within a COMPAQ-386 portable computer.

Figure 3 also shows a diagram of the apparatus used to generate the leak in the pipeline. The flow rate of the leak was controlled by the 0 to 170 kPa (0 to 25 psi)

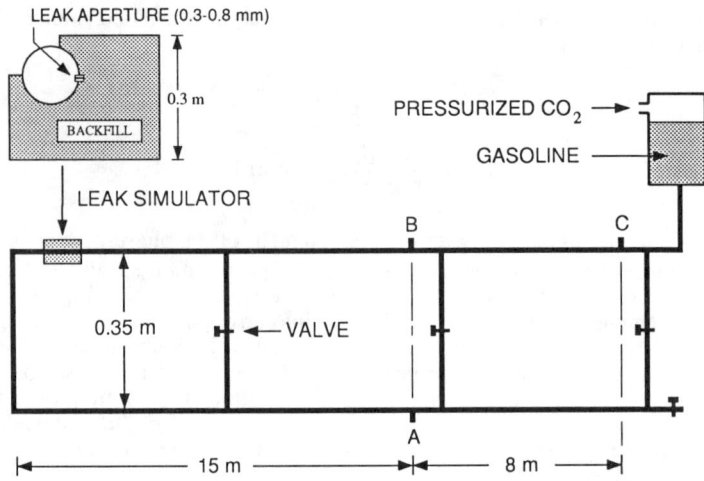

FIG. 3 – Diagram of the pressurized petroleum pipeline at the UST Test Apparatus. Pressurized CO_2 is used to generate static line pressure. A cross section of the leak simulator is also shown.

static line pressure and the diameter of the aperture through which the product was allowed to leak. Leak apertures between 0.4 and 0.7 mm were introduced into the pipeline via carburetor jets in order to avoid the difficulty of drilling small-diameter holes through the steel wall of the pipeline. The range of flow rates generated during the experiments was between 2 and 20 L/h (0.5 and 5.0 gal/h). The backfill materials used in the experiments were fine-grain sand and pea gravel.

Three types of acoustic measurements (calibration, leak-on, and leak-off) were performed for each combination of line pressure, hole diameter, and backfill material. The calibration signal was produced by breaking a pencil lead on the pipe surface near the location of the simulated leak. The relative arrival times of this impulsive signal at the three transducer locations were used to verify that the sensors and data acquisition system were operating properly. After the initiation of the leak, approximately eight leak-on measurements 1.7 s in duration were recorded at one-minute intervals. The leak-on measurements were bracketed by a pair of recordings obtained under leak-off conditions.

DATA

The raw data consist of time series of acoustic leak signals and ambient noise sampled simultaneously by three sensors. The first step toward applying a leak location algorithm to the raw data is to view the data in three forms: (1) time series, (2) power spectral density, and (3) complex coherence. Viewed in the time domain, the leak-on/leak-off data (i.e., time series) provide convincing evidence that an acoustic leak signal exists and is detectable over the dimensions of the pipeline. In addition, the time series reveal something of the character of the leak signal. However, the time series alone offer no clues as to the location of the leak or the types of processing required to perform a source location estimate. The distribution, with respect to frequency, of acoustic energy emitted by the leak and the way in which this energy is propagated from source to sensor is revealed by viewing the data in the frequency

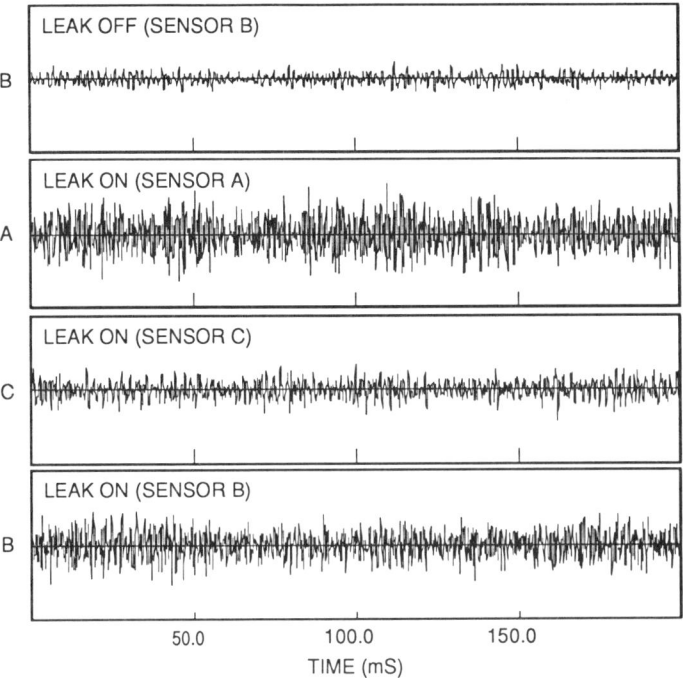

FIG. 4 – Time series of acoustic leak signals generated by a 11.4 L/h leak. Sample rate is 10 kHz. A no-leak time series recorded by sensor B is shown for reference.

domain (i.e., power spectra and complex coherence).

Time series of acoustic leak signals generated by a 11.4-L/h (3.0-gal/h) gasoline leak into a sand backfill are shown in Figure 4; a time series recorded under no-leak conditions by one of the sensors is shown for reference. Aside from an anti-alias filter applied to the analog signals prior to digitization, the time series presented here represent unfiltered data. Due to the low level of ambient acoustic noise associated with the underground pipeline, the fluctuations observed in the leak-off time series of Figure 4 are determined largely by electronic noise within the amplifiers. The distance between the simulated leak and sensors A and B is approximately 15 m; sensor C is located approximately 23 m from the leak. The line pressure used in this experiment was 100 kPa (15 psi) and the hole diameter was 0.7-mm. Two important observations should be made regarding the time series of Figure 4: (1) a comparison of the leak-on and leak-off measurements clearly shows that the leak is detectable, and (2) the relative arrival time of the leak signal at the different sensor locations cannot be obtained through inspection of the time series. The continuous nature of the acoustic leak signal requires that some type of signal processing be applied to the leak signal time series in order that the relative arrival times, and hence the location of the leak, can be estimated.

The strength of the acoustic signal produced by a leak in a buried pipeline is proportional to the flow rate, for a hole of a given diameter. Estimates of the signal-to-noise ratio (SNR) for pipeline leaks into a sand backfill at flow rates of 11.4, 5.7, 3.8, and 1.9 L/h are shown in Figures 5a-d. The hole diameters and line pressures used to establish the flow rates were 0.7 mm at 100 kPa (15 psi), 0.5 mm at 100 kPa (15 psi), 0.4 mm at 100 kPa (15 psi), and 0.4 mm at 35 kPa (5 psi), respectively. The

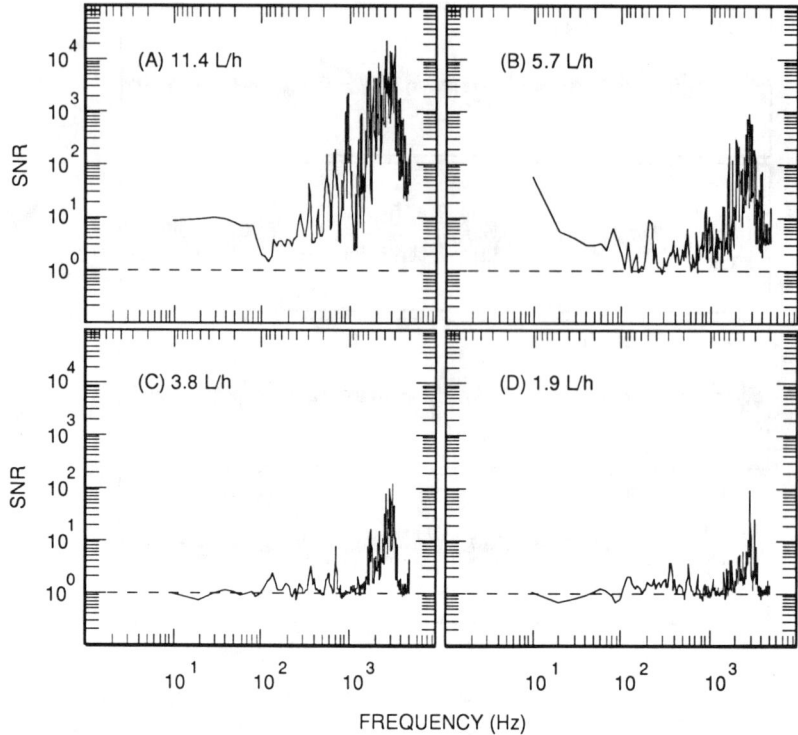

FIG. 5 – Signal-to-noise ratio (SNR) for pipeline leaks at flow rates of 11.4 L/h (A), 5.7 L/h (B), 3.8 L/h (C), and 1.9 L/h (D). Dashed line indicates SNR=1.

SNR at each flow rate was obtained by dividing the power spectral density computed with the leak present by a similar spectrum computed with no leak present. The power spectra for each of the three individual sensors, computed using 31 overlapping, 1024-point FFT segments, were averaged together prior to computing the SNR. The time series used were 1.7 s in duration and were sampled at a frequency of 10 kHz. The SNR spectra show that the energy associated with the acoustic leak signal is not equally distributed over the 1- to 5000-Hz sampling bandwidth, but is instead concentrated within a relatively narrow 1- to 4-kHz frequency band. The frequency domain representation of acoustic data offers a means by which the location algorithm can separate useful information concerning the leak from unwanted noise.

Figure 6a shows the coherence amplitude as a function of frequency for acoustic leak signals received by sensors bracketing a 5.7-L/h leak. The sensor separation is 38 m. The coherence plot represents an ensemble average of 15 overlapping, 1024-point segments, each individually detrended and weighted with a cosine bell prior to Fourier transforming. Statistically significant coherence (as indicated by the 95% confidence level) is observed primarily within the frequency bands 0.9 to 1.2 kHz and 2.0 to 4.0 kHz. It should be noted that within both of these frequency bands γ^2 is not statistically significant at all Fourier frequencies. Figure 6b shows the coherence amplitude for acoustic leak signals received by sensors bracketing a leak through a 0.4-mm-diameter hole pressurized to 35 kPa (5 psi); the flow rate is 1.9 L/h and the sensor separation is

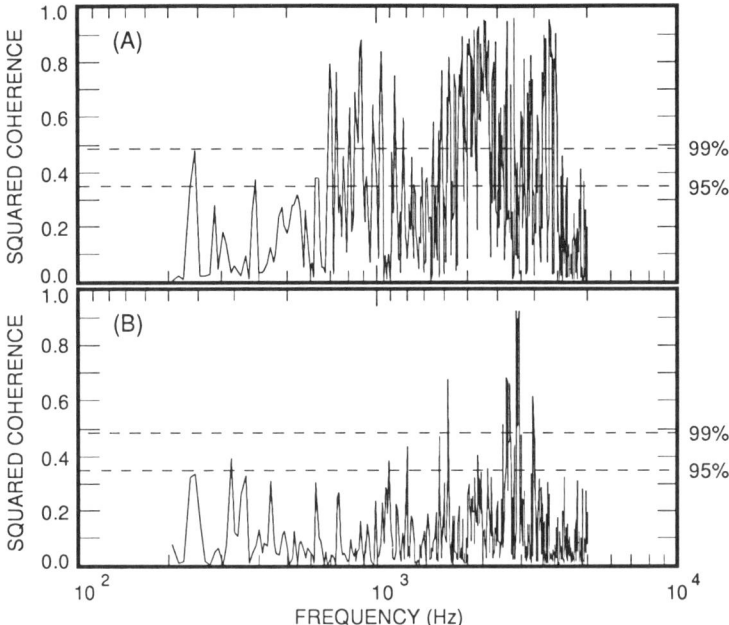

FIG. 6 – Coherence amplitude as a function of frequency for acoustic signals generated by 5.7 L/h (A) and 1.9 L/h (B) leaks. Sensor separation is 38 m (A) and 30 m (B). Dashed lines indicate 95% and 99% levels of statistical significance.

30 m. As the line pressure is reduced, the frequency band within which signal similarity is maintained is narrowed considerably.

LOCATION RESULTS

Table 1 summarizes the results of leak location and wave speed estimates for flow rates of 11.4, 5.7, and 3.8 L/h. Leak location estimates are reported as a difference between the computed and actual location. Application of a leak location algorithm based upon the technique of coherence function analysis resulted in mean differences between predicted and actual leak locations of 8.7 cm (11.4 L/h), 3.6 cm (5.7 L/h), and -11.6 cm (3.8 L/h). Standard deviations of the location estimates were 26.1 cm (11.4 L/h), 26.3 cm (5.7 L/h), and 39.1 cm (3.8 L/h). The mean propagation speed was 915 m/s with a standard deviation of 146 m/s.

The procedure used to estimate the leak location and wave speed for a given set of time series is as follows: (1) compute the coherence function between the three sensor pairs (i.e., A-B, A-C, and B-C), (2) identify frequency bands of at least 100-Hz width for which the coherence amplitude exceeds the 95% level of statistical significance, (3) unwrap the coherence phase within these frequency bands, (4) compute the linear regression lines through each of the three $\phi(f)$ curves, and (5) apply Eqs. (5) through (7), using the known sensor positions and the computed regression slopes.

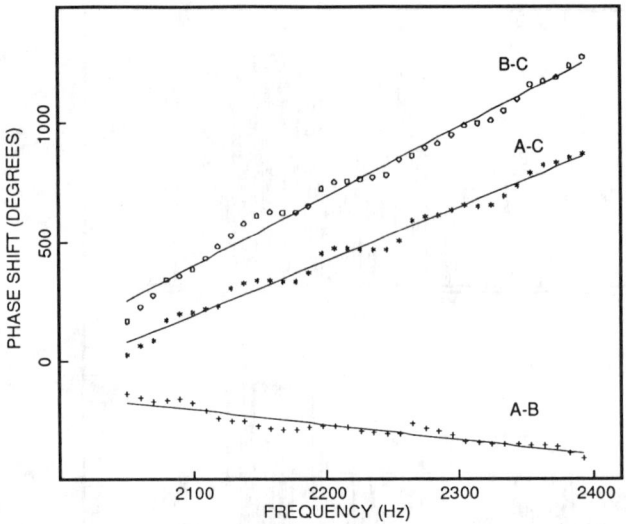

FIG. 7 – Unwrapped coherence phase between 2.0 and 2.5 kHz for sensor pairs A-B, A-C, and B-C of Figure 3. Least-squares regression lines through actual data points are included. The flow rate is 11.4 L/h.

TABLE 1-- Leak location and propagation speed measurements.

Flow Rate (L/h)	D (mm)	P (kPa)	Δf [1] (Hz)	Mean Error (AB) [2] (cm)	Std. Dev. (AB) [2] (cm)	Mean Error (AC) [3] (cm)	Std. Dev. (AC) [3] (cm)	V (m/s)	σ_V (m/s)	N_L [4]
11.4	0.7	140	2100-2400	8.6	16.4	-2.4	23.7	1048	37	25
11.4	0.7	140	3800-4050	18.7	29.9	14.2	31.8	917	89	18
5.7	0.5	140	2100-2400	14.4	15.8	15.8	14.9	930	136	23
5.7	0.5	140	3800-4050	-5.8	19.8	-12.2	20.4	775	81	15
3.8	0.5	76	3800-4050	-2.5	47.9	-20.7	28.1	715	150	8

[1] Location algorithm analysis frequency band
[2] A-B used as bracketing sensors
[3] A-C used as bracketing sensors
[4] Number of independent location estimates

If the coherence amplitude is statistically significant for each Fourier component within a given frequency band, a simple phase-unwrapping procedure can be applied to the coherence phase. Figure 7 shows the unwrapped phase differences between sensor pairs A-B, A-C, and B-C for the frequency band 2.0 to 2.5 kHz. The flow rate used in this experiment was 11.4 L/h. Included in this plot are least-squares regression lines through the actual data points corresponding to each sensor pair. The criterion for the inclusion of a phase measurement in the estimation of leak location and wave speed is that the coherence amplitude exceed the 95% level of statistical significance for each of

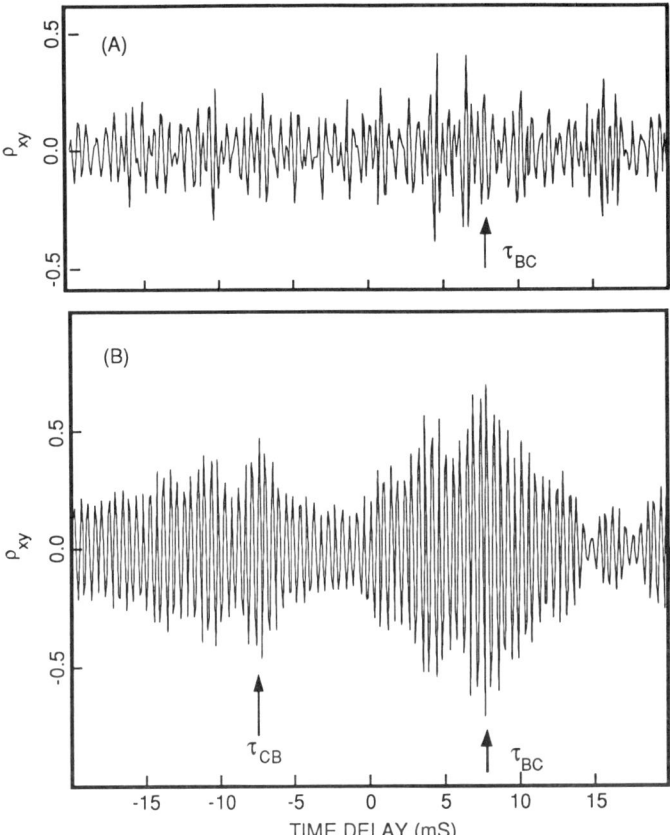

FIG. 8 – Normalized cross-correlation coefficient as a function of time delay between time series recorded by sensors B and C. The time series were bandpass-filtered between 1.0 and 4.0 kHz (A) and 2.0 and 2.5 kHz (B) prior to computing ρ_{xy}. τ_{BC} and τ_{CB} represent predicted time delays for primary and reflected acoustic waves propagating at V=1000 m/s.

the three sensor pairs at a given Fourier frequency. The regression slopes of Figure 7 can be used to calculate the time delays between signals received by the three sensor pairs. The measured $d\phi/df$ values of -0.47 °/Hz (A-B), 2.07 °/Hz (A-C), and 2.55 °/Hz (B-C) correspond to time delays of -1.3, 5.8, and 7.1 ms, respectively.

An alternative method of extracting the time delays from the time series is to apply the technique of cross-correlation. Figure 8a shows the normalized cross-correlation coefficient as a function of lag time, $\rho_{xy}(\tau)$, between the time series B and C used in Figure 7. The time series were bandpass filtered in order to isolate the high-SNR, 1.0- to 4.0-kHz portion of the leak signal spectrum prior to computing the correlation coefficient. Without the detailed knowledge of the distribution of leak signal energy provided by the coherence function, correlation analysis fails to give an accurate measurement of the time delay between leak signals received by sensors B and C.

Figure 8b shows the correlation coefficient computed between B and C time series in which the data are bandpass filtered from 2.0 to 2.5 kHz. Within the high-coherence interval used to generate the phase curves of Figure 7, correlation analysis and coherence function analysis result in approximately equal estimates of the time delay. Although this result suggests that the two techniques for measuring time delays are equivalent, accurate correlation analysis requires *a priori* knowledge of the frequency bands within which the acoustic leak signal is strong and composed of linearly propagating waves. Coherence function analysis identifies frequency bands for which the SNR is high (through the coherence amplitude) and for which the phase behavior is appropriate for leak location (through the coherence phase).

LEAK SIGNAL PROPAGATION

The analysis of acoustic data from pipelines is complicated by the presence of multi-path and multi-mode wave propagation. Multi-path signals are produced by reflections within the complex pipeline geometry or by signal leakage, across the connecting arms, from one main branch of the pipeline to the other (see Figure 3). Multi-mode wave propagation results from the excitation, by the leak flowfield, of wave motion in different materials (e.g., gasoline and steel), or of waves in the same material that propagate at different speeds (e.g., longitudinal and transverse waves). While the analysis presented above suggests that the acoustic leak signal is dominated by a single propagation mode that traverses a single path from leak to sensor, experimental data and simple simulations show that the effects of multi-path and multi-mode propagation are detectable.

The reflective nature of the pipeline is illustrated by the cross-correlation plot shown in Figure 8b. The primary ρ_{xy} peak, which occurs at the lag time $\tau \approx 7$ ms, corresponds to signals propagating in the direction from sensor B to sensor C at speed $c \approx 1000$ m/s. A secondary peak, which occurs at the lag time $\tau \approx -7$ ms, is consistent with reflection signals propagating at the same speed, but in the opposite direction.

Energy propagation along the pipeline results from the excitation of three types of wave motion by the leak flowfield: (1) transverse waves propagating in steel, (2) longitudinal waves propagating in steel, and (3) longitudinal waves propagating within the product contained in the pipeline. The nominal propagation speeds for each type of wave motion are 6000 m/s (longitudinal, steel), 3000 m/s (transverse, steel), and 1200 m/s (longitudinal, gasoline). The similarity between the measured wave speed (≈ 1000 m/s) and the speed of acoustic waves in gasoline suggests that in the far field of the leak, the sensors respond primarily to longitudinal waves propagating through the product. These longitudinal waves are sensed indirectly through stresses induced in the steel in response to the fluctuating pressure field within the pipe. If other forms of wave motion are produced by the leak and are detectable, the phase measurements, and thus the location estimates, will be degraded.

The detectability of longitudinal waves propagating in steel was investigated through a calibration test in which an impulsive signal was generated by breaking a pencil lead near the leak location. Figure 9 shows time series of the calibration impulse received by sensors B and C. The measured time delay (1.2 ms) and sensor separation (7.5 m) yield a propagation speed of 6250 m/s for the leading edge of the impulse. This speed is consistent with the nominal value of 6000 m/s for longitudinal waves propagated within steel. While the calibration data do not indicate the degree to which the longitudinal wave mode in steel is excited by the leak flowfield, it does show that such waves, if emitted by the leak, will be detected by sensors mounted externally on the pipeline wall.

The excitation of transverse waves by the leak flowfield, and their detectability, were investigated through a series of experiments in which CO_2, rather than gasoline,

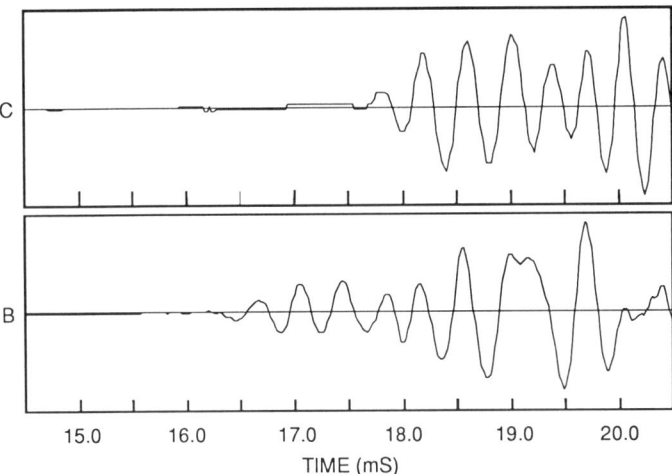

FIG. 9 – Time series of impulsive calibration signals recorded by sensors B and C. The estimated propagation speed (6250 m/s) is consistent with the nominal speed of sound in steel.

was used as the product. Time series of acoustic leak signals produced by the flow of CO_2 through a 0.7-mm-diameter hole under 100-kPa line pressure were recorded by the three-element sensor array. Application of Eq. (7) to the phase plot corresponding to sensor pair B-C yielded a propagation speed of approximately 2400 m/s. Two important observations should be noted regarding this experiment: (1) the measured wave speed is similar to the nominal value for freely propagating transverse waves in steel, and (2) the measured wave speed is much higher than the speed of acoustic waves propagated in CO_2 ($c \approx 270$ m/s). The SNR of the CO_2 leak was approximately 15 dB less than the SNR recorded in the presence of a gasoline leak at the same line pressure and hole diameter. Two conclusions may be drawn from these measurements: (1) freely propagating transverse waves are produced by the leak and are detectable in the far field, and (2) the coupling between acoustic waves in the product and stresses induced in the surrounding pipeline is a function of the product contained within the pipeline. Liquid leaks appear to be sensed primarily through energetic, low-velocity acoustic waves, while gas leaks are sensed via less energetic, high-velocity transverse waves propagating in the steel.

The effect of multi-path and multi-mode wave propagation can also be observed in the coherence phase. Figure 10 shows a plot of the phase shift between sensors B and C in which the linear trend has been removed. The residual phase shift is dominated by a non-random, periodic oscillation that occurs at intervals of approximately 50 Hz with an average amplitude of 40°. If the signal received at each sensor is represented as a summation of a direct-path signal propagating at the observed wave speed and contaminating signals caused by multi-path and multi-mode propagation, an estimate can be made of the fraction of total energy received via the contaminating signals. A simple simulation in which approximately 15% of the total received energy was propagated by multi-path and multi-mode waves produced residual phase shifts comparable to those observed in the data.

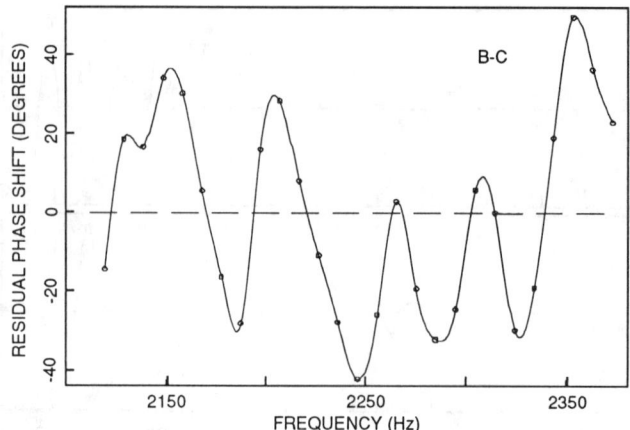

FIG. 10 – Unwrapped coherence phase between 2.1 and 2.4 kHz for sensor pair B-C in which the linear trend has been removed. The flow rate is 11.4 L/h.

PHASE UNWRAPPING

Accurate source location requires that the location algorithm distinguish between useful information provided by the leak signal, and ambient or system noise. The continuous nature of the acoustic leak signal further requires that the separation of signal from noise take place in the frequency domain, through coherence function analysis, rather than in the time domain. It has been demonstrated that source location through cross-correlation analysis is not accurate when applied to wide frequency bands (e.g., the 1.0 to 4.0 kHz frequency band used in Figure 8a). While the location estimates given in Table 1 are based upon the successful application of coherence function analysis to relatively narrow frequency bands (100 to 500 Hz), the possibility exists that a similar location algorithm may be applied to frequency bands of arbitrary width.

When the relative separation between a pair of sensors is large compared to the wavelength of the received signals, some form of phase-unwrapping algorithm must be applied in order to measure $d\phi/df$ over a wide range of frequencies. Such unwrapping algorithms are easily implemented, provided that the coherence phase is reliably measured (i.e., that the coherence amplitude is high) at many frequencies within the desired band. As the distribution of reliable phase estimates within a frequency band becomes more sparse, the ability to simply unwrap the phase is diminished, and the information provided by the phase measurements must be discarded. If the leak location and propagation speed of acoustic waves are known, the correspondence between measured and predicted phase shifts can be viewed over an arbitrarily wide frequency band.

Figures 11 and 12 show the unwrapped phase shift between sensors A-B, A-C, and B-C, in which the unknown multiples of 360° required to unwrap the phase were computed from the predicted $\phi(f)$ lines (shown as solid lines in the figures). Reliable phase measurements (indicated by markers in the plots) correspond to coherence amplitudes that exceed the 95% level of statistical significance; the flow rates are 11.4 L/h (Figure 11) and 5.7 L/h (Figure 12). The frequency distribution of reliable phase measurements for the 11.4-L/h data is such that all of the information contained in the 2.0- to 4.0-kHz band can be used in the location estimate if a straightforward

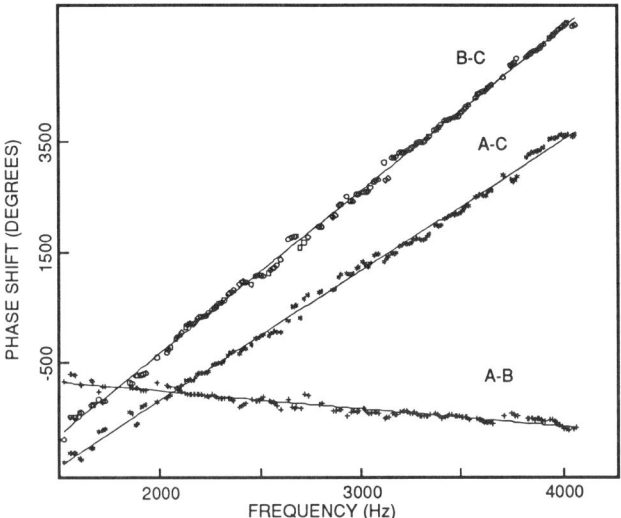

FIG. 11 – Unwrapped coherence phase between 1.5 and 4.5 kHz for sensor pairs A-B, A-C, and B-C. Solid lines indicate predicted coherence phase for linearly propagating plane waves based upon known leak location and propagation speed. Flow rate is 11.4 L/h.

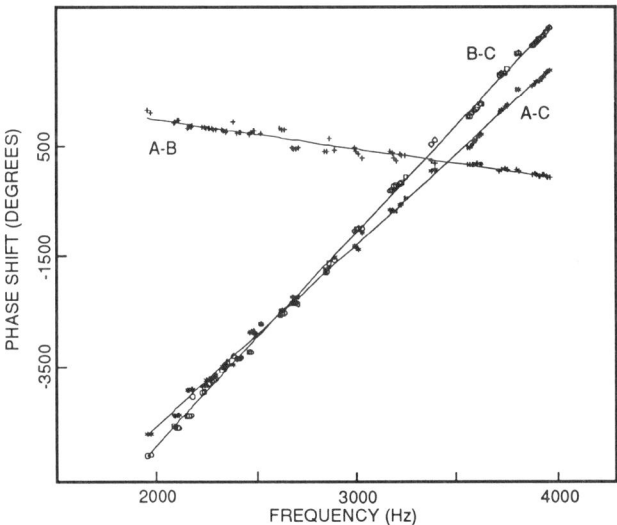

FIG. 12 – Unwrapped coherence phase between 1.5 and 4.5 kHz for sensor pairs A-B, A-C, and B-C. Solid lines indicate predicted coherence phase for linearly propagating plane waves based upon known leak location and propagation speed. Flow rate is 5.7 L/h.

phase-unwrapping algorithm is implemented. As the flow rate is reduced, however, the simple unwrapping algorithm works only within a small number of narrow frequency bands (e.g., 2.2 to 2.5 kHz, and 3.7 to 4.0 kHz in Figure 12). The similarity between the measured and predicted phase shift outside of these narrow bands suggests that a more robust unwrapping algorithm may be capable of exploiting a greater fraction of the available phase information for the purpose of leak location.

CONCLUSIONS

Passive acoustic measurements, combined with advanced signal processing techniques based on coherence analysis, offer a promising method for the location of small leaks in the pressurized petroleum pipelines found at retail service stations and industrial storage facilities. While the results presented in this work represent a significant improvement over previous pipeline leak location efforts, a better understanding of the underlying physics of pipeline acoustics, including the propagation modes and source mechanisms of the acoustic leak signal, will help optimize the location algorithm and the instrumentation.

Experiments were conducted on a 2-in.-diameter underground pipeline at the UST Test Apparatus in which three acoustic sensors separated by a maximum distance of 38 m were used to monitor signals produced by 11.4-, 5.7-, and 3.8-L/h gasoline leaks. Application of a leak location algorithm based upon the technique of coherence function analysis resulted in mean differences of approximately 10 cm between predicted and actual leak locations. Standard deviations of the location estimates were approximately 30 cm.

The full capability of the location algorithm was not evaluated in these tests. The smallest hole used to generate a leak in the experiments was 0.4 mm. At a line pressure of 140 kPa (20 psi) this resulted in a leak rate of 3.8 L/h (1 gal/h). Additional experiments need to be performed with smaller holes and at higher line pressures (150 to 350 kPa) to determine the minimum leak rate that can be reliably located.

Spectra computed from leak-on and leak-off time series indicate that the majority of acoustic energy received in the far field of the leak is concentrated in a frequency band from 1 to 4 kHz. Energy propagation from leak to sensor was observed via three forms of wave motion: longitudinal waves in the product, transverse waves in the steel, and longitudinal waves in the steel. Though each of these propagation modes is believed to contribute to the overall received signal, longitudinal wave motion in the product was clearly the dominant propagation mode for liquid-filled pipelines. The effects of multiple-mode wave propagation and the reflection of acoustic signals within the pipeline were observed as non-random fluctuations in the measured phase difference between sensor pairs.

The SNR was observed to be generally high within the entire 1.0- to 4.0-kHz frequency band; however, continuous regions of high coherence appropriate for source location were typically 100 to 500 Hz in width. Several data sets recorded in the presence of the 11.4-L/h leak exhibited high coherence over a 2-kHz bandwidth. Location estimates obtained by means of cross-correlation showed that without the detailed knowledge of signal similarity provided by the coherence function, cross-correlation analysis cannot locate small leaks with acceptable accuracy. The observed correspondence between measured and predicted phase shifts within the 1.0- to 4.0-kHz analysis band demonstrates the need to develop a more sophisticated location algorithm such that a greater fraction of the information contained in coherent leak signals may be processed.

Buried pipelines provide a generally quiet ambient environment in which to perform acoustic measurements. Since the SNR for a given leak largely determines the ability of a passive acoustic system to locate the leak, the system noise level should be determined by ambient acoustic noise, rather than electronic noise. The combination of sensors and preamplifiers used in these experiments was incapable of resolving the low levels of ambient acoustic noise associated with the pipeline. Improved system performance may be attained through the use of transducers with greater sensitivity in the low frequency range (1 to 10 kHz) and low-noise preamplifiers.

ACKNOWLEDGEMENTS

This work was funded by the U.S. EPA under contract No. 68-03-3409. The authors gratefully acknowledge CTI, Inc., for the loan of the acoustic sensing equipment used in the experiments.

REFERENCES

[1] D.S. Kupperman, T.N. Claytor, T. Mathieson, and D. Prine, "Leak Detection Technology for Reactor Primary Systems," Nuclear Safety, Vol. 28 (April-June 1987).

[2] D.S. Kupperman and D.E. Karvelas, "Acoustic Leak Detection for District Heating Systems," Technical Report No. ANL-87-60, Argonne National Laboratory, Argonne, Illinois (February 1988).

[3] P.R. Roth, "Effective Measurements Using Digital Signal Analysis," IEEE Spectrum, Vol. 8 (April 1971).

[4] E.G. Eckert and J.W. Maresca, Jr.,"Location of Leaks in Pressurized Petroleum Pipelines by Means of Passive-Acoustic Sensing Methods," EPA Contract 68-03-3409, Risk Reduction Engineering Laboratory, U.S. Environmental Protection Agency, Edison, New Jersey (1991).

[5] J.S. Bendat and A.G. Piersol, Engineering Applications of Correlation and Spectral Analysis (New York: John Wiley & Sons, 1980).

External Monitoring

Robert P. Schreiber[1] and Myron S. Rosenberg[2]

ANALYSIS OF UST LEAK VAPOR DIFFUSION AND LIQUID BUILD-UP

REFERENCE: Schreiber, R. P. and Rosenberg, M. S., "Analysis of UST Leak Vapor Diffusion and Liquid Build-Up," Leak Detection for Underground Storage Tanks, ASTM STP 1161, Philip B. Durgin and Thomas M. Young, Eds., American Society for Testing and Materials, Philadelphia, 1993.

ABSTRACT: The need for improved leak detection and corrective action has prompted research into the movement of hydrocarbon vapors from leaking underground storage tanks (USTs) as well as the build-up of liquid hydrocarbon on the water table. This research has included the development of two evaluation techniques, one for simulating vapor diffusion from an UST leak and another for simulating the mounding of leaked hydrocarbon liquid. Both techniques are designed to produce approximate estimates of hydrocarbon movement and build-up, and as such are intended to be used in the early stages of site remediation planning and monitoring. The result of the research is a set of response curves and analytical techniques that can be used in designing monitoring systems and in performing site clean-ups.

KEY WORDS: UST, leak detection, leak monitoring, soil vapor, NAPL, floating product.

OVERVIEW

The need for improved leak detection and corrective action has prompted research into the movement of hydrocarbon vapors from leaking USTs as well as the build-up of liquid hydrocarbon on the water table. This research has resulted in the development of response curves and analytical techniques that can be used in designing monitoring systems and in performing site clean-ups.

---

[1]Associate, Camp Dresser & McKee Inc., 10 Cambridge Center, Cambridge, MA 02142.

[2]Vice President, Camp Dresser & McKee Inc., 10 Cambridge Center, Cambridge, MA 02142.

To evaluate vapor movement in relation to soil characteristics and UST leak locations, a three-dimensional finite element model of vapor diffusion was developed. This model simulated diffusion-limited vapor transport in a hypothetical UST excavation zone and surrounding native soils. A synthetic gasoline was the hydrocarbon liquid stored in the UST, for which physicochemical properties were derived. Simulations of a point leak were performed, with varying temperatures, surface boundary conditions, and soil characteristics including porosity and moisture content. The model results were expressed as time histories of vapor concentrations measured at various points in the subsurface. These were converted to response curves, plotting amount of leaked product versus the time to reach detectable vapor levels. The modeling showed that external vapor sensors can function as good "early-warning" devices in many situations.

For analyzing the build-up of liquid hydrocarbons on the water table, an analytical solution was derived based on a method for predicting the shape of the interface between fluids of different densities. Application of this technique relies upon the measurement of hydrocarbon thickness at two or more monitoring wells. It is applicable to sites with medium- or coarse-grained materials that exhibit small capillary effects; it is not applicable to sites with silts or clays. This analytic approach yields an estimate of the amount of leaked UST product. It can also be used to predict the thickness of the hydrocarbon "lens" on the water table, if the leaked amount is known. The technique, therefore, is especially useful during the early stages of corrective action for making preliminary estimates of hydrocarbon thickness, extent, or leaked volume.

This paper describes these two evaluation techniques, and shows how they can be used to improve the design of monitoring networks and to enhance the process of site clean-up. The remainder of this paper is a summary of two prior papers that have appeared elsewhere (Schreiber, et al., 1988; Levy, et al., 1990) [1], [2]. In general, the prior papers contain more details on the simulation and analysis methods, as well as mathematical derivations and explanations. This paper summarizes the essential details of these derivations, with an emphasis on the application of the techniques.

VAPOR DIFFUSION SIMULATION

Simulation modeling of subsurface vapor transport from a leaking UST was performed to provide guidance to the U.S. Environmental Protection Agency (EPA) on the use of vapor detectors. This modeling was conceived as part of an overall effort by the research branch of EPA to evaluate systematically the "promise" of vapor sensors. Specifically, it was anticipated that vapor detectors could act as "early warning" devices, and, in combination with other techniques, could provide a good level of protection. How well vapor sensors could perform has not been quantified. The modeling described below was an important step in taking vapor sensors from the realm of "promising" devices to the status of quantified and tested detectors as a key component in an UST leak detection system. This work was reported in detail in Camp Dresser & McKee (1988)[3].

The intent of the work was to provide quantitative estimates of how quickly vapor sensors might respond to a leak. Emphasis was placed on obtaining order-of-magnitude estimates, rather than analyzing all of the complexities of vapor transport in detail. A pragmatic approach was adopted in which simplifying, yet justifiable assumptions were made concerning vapor transport phenomena, site-specific conditions, tank contents, and leakage characteristics.

MODELING ASSUMPTIONS AND APPROACH

The first simplifying assumption was that leakage from a single tank would be simulated. Figure 1 shows the simulation tank, which was assumed to be cylindrical and located in a box-shaped excavation zone, surrounded by at least 2 feet (0.61 meters) of backfill material. The dimensions of the UST were selected to represent a typical tank, with a diameter of 6 feet (1.83 meters) and a length of 12 feet (3.66 meters), storing about 2 500 gallons (9 462 liters). Although the water table was fixed in the modeling, the soil moisture content was a key parameter that was varied to evaluate its effect on sensor responses. Figure 2 shows the hypothetical detector locations used in the simulation modeling.

Because the majority of USTs are at gasoline stations, the simulated tank was assumed to store gasoline. Table 1 lists the blend formulated for the modeling. It is not a commercially-available gasoline, but is rather a synthetic blend of common unleaded gasoline components.

Estimating and evaluating the physicochemical properties of the gasoline blend and its individual components was an essential part of the vapor modeling. Not only do the properties provide the basis for defining the vaporization at the leak, but they also lend insight into how vapor sensors can best respond to gasoline leaks. For example, the gasoline vapors that will most likely first reach vapor sensors are not benzene, toluene, or xylenes (BTX), but instead are isopentane and isobutane, because the latter two are found in higher percentages in most gasoline blends, and they have a higher vapor potential.

Vapor diffusion was the sole mode of transport simulated. Diffusion coefficients for vapors in soils were estimated using measured air diffusion coefficients and theoretical/empirical formulas for soil tortuosity and soil-vapor diffusion, based on soil porosity and moisture content (Millington and Quirk, 1961)[4]. The diffusion simulation was based on the observation that the rates of diffusion and vapor flux are proportional to the vapor concentration gradient, or Fick's Laws.

76 LEAK DETECTION FOR UNDERGROUND STORAGE TANKS

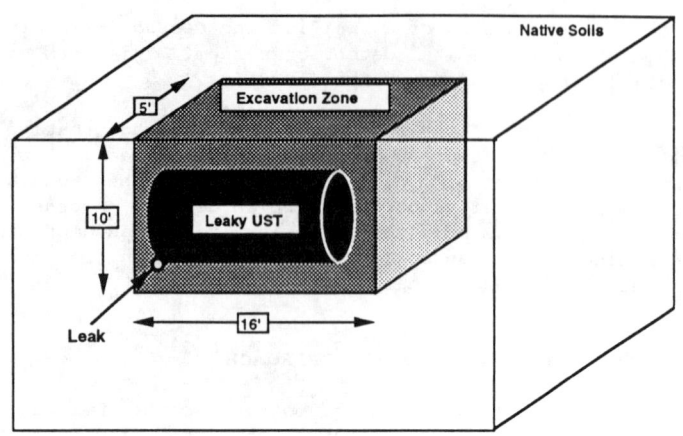

Figure 1
Vapor Diffusion Model Geometry,
Cross-Section Through UST
(1 m = 3.28 ft)

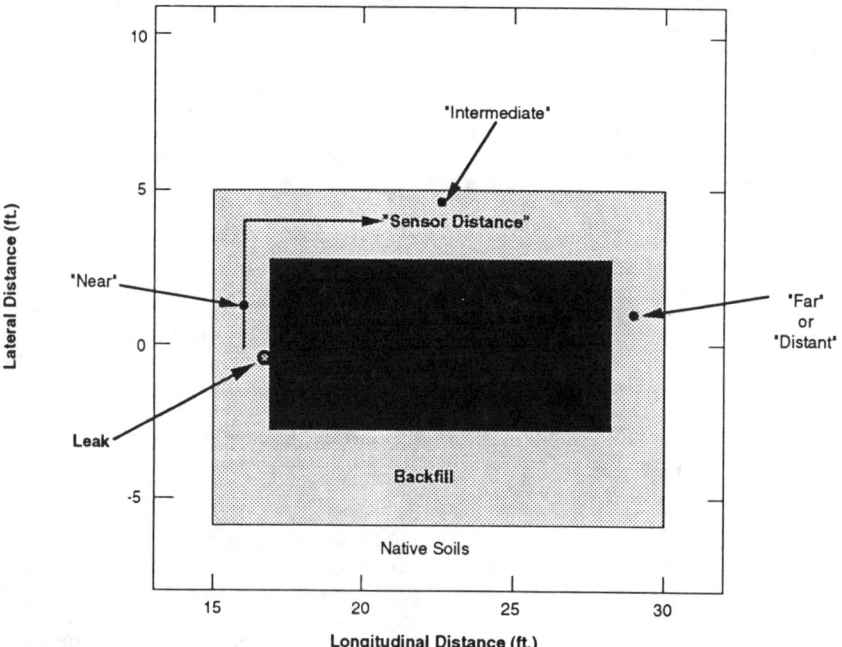

Figure 2
"Sensor Distance" and "Sensor Locations" in Plan View
(1 m = 3.28 ft)

TABLE 1--Synthetic Gasoline Blend

Chemical Components of Blend

Chemical	Percent of Blend	
	Liquid	Vapor
Benzene	3.0	0.9
Ethylbenzene	2.0	<0.1
1,4-Diethylbenzene	5.0	<0.1
1,3,5-Trimethylbenzene	5.0	<0.1
Isobutane	2.0	28.5
n-Butane	1.0	9.6
C_{12}-aliphatic	10.0	<0.1
n-Heptane	1.5	0.2
Cyclohexane	3.0	0.9
Methylcyclohexane	1.0	0.1
n-Hexane	9.0	4.0
2-Methylhexane	5.0	0.8
2,4-Dimethylhexane	8.0	0.5
2,2,4-Trimethylhexane	2.0	<0.1
2,2,5,5-Tetramethylbenzene	1.5	<0.1
1-Hexene	1.5	0.9
n-Octane	1.0	<0.1
Isopentane	14.0	38.2
n-Pentane	3.0	5.9
2-methylpentane	8.0	5.1
1-Pentane	1.5	3.9
Toluene	5.0	0.4
m-xylene	7.0	0.1

The leak source was assumed to be a "point" at the bottom of one end of the tank. The rate of leakage was governed by the diffusion of vapors away from the leak. The leak started at simulation time zero, going from a zero concentration immediately to the 100 percent concentration for the simulated gasoline blend. This was termed "base case" leakage because only vapors were simulated to escape at the "point source" leak. Under the "base case" assumption, gasoline leakage is exactly balanced by diffusion.

Vapor-water partitioning was investigated, and an equivalent retardation factor was applied to reduce the diffusion coefficient. It was determined, however, that this process would not have a significant influence on overall transport relative to diffusion. This is because such retardation would affect only those chemicals with high Henry's Law constant values. Such chemicals constitute a small percentage of gasoline vapors.

SIMULATION MODEL

A computer program, DYNFLOW (Camp Dresser and McKee, 1984)[5], was used to perform the vapor transport simulations. It simulates three-dimensional groundwater flow, under steady-state or transient (dynamic) conditions. DYNFLOW was used because the governing equation for vapor diffusion is identical to the one for confined groundwater flow, with the effective soil vapor diffusion coefficient replacing "aquifer diffusivity," and vapor concentration replacing piezometric head. A complete development and discussion of the mathematical derivation of the vapor diffusion governing equations is provided in Camp Dresser & McKee (1988)[3] and Levy, et al (1988)[6].

To verify the applicability and accuracy of DYNFLOW, a test case was simulated for which an analytical solution was available. One-dimensional diffusion from a constant point source was simulated using a long, rectangular-shaped model grid, with the source as 100 percent concentration at one end of the grid. Close agreement between theory and model was achieved, in terms of concentration profiles and vapor fluxes over time.

The modeling assumptions and computer program were validated by simulating vapor transport experiments conducted by Dr. Richard Johnson of the Oregon Graduate Center (OGC). This validation, which is described in Schreiber, et al. (1988)[7], was performed in cooperation with the University of Nevada at Las Vegas (Weber, 1988)[8].

The OGC experiments were conducted in a large, in-ground, sand-filled tank. Two rounds of experiments were run: the first was intended to create "diffusion only" transport with low volumes of injected contaminants; the second round was designed to develop gravity-induced advection by injecting much larger contaminant volumes. One quadrant of the OGC "sand box" was simulated using a diffusion coefficient and retardation factors computed as described above.

Agreement between the model results and experimental "diffusion only" data was generally good. The only exception was the comparison to toluene: the model predicted more vapor-water partitioning than what actually occurred. Because over-predicting retardation is "conservative," it was concluded that the modeling approach is a reasonable one.

MODEL APPLICATION

The single tank and its surrounding excavation zone and native soils, as described above, were transformed into a three-dimensional numerical model consisting of a "one-half" symmetric cut of the tank and surrounding soils and backfill. The sides of the native soil portion of the model were simulated as fixed zero-concentration boundaries. The sides were placed far enough from the tank so that sensors in the excavation zone would reach detection limits before significant amounts of vapors could reach the edges. The "no-flow" boundary at the bottom of the model represented a water table that is 24 feet (7.31 meters) from the ground surface.

To display the simulation results with detector network design in mind, "sensor distance" and "sensor locations" were defined (Figure 2). At the three "sensor locations," model results were displayed for both "deep" sensors, 1 foot (0.30 meters) above the bottom of the excavation zone, and "shallow" sensors 2 feet (0.61 meters) below the surface.

Six combinations of backfill and native soil properties were simulated (Table 2). The combinations were formulated to show the difference in sensor response and "base case" leakage rates to changes in the porosity and moisture content of the backfill and the native soil. All simulations reported herein were for a paved surface, impervious to vapors.

TABLE 2--Simulation Runs

Simulation Run Number	Backfill Material	Native Soil
1	Dry Gravel	Dry Gravel
2	Dry Gravel	Dry Silty Sand
3	Moist Sand & Gravel	Moist Sand & Gravel
4	Moist Sand & Gravel	Wet Silty Sand
5	Wet Sand & Gravel	Wet Sand & Gravel
6	Wet Sand & Gravel	Wet Clay

VAPOR MODELING RESULTS

The most significant finding of the vapor modeling has been that, under "dry" backfill conditions, a vapor sensor responding to about 500 parts per million (ppm) of total organic hydrocarbons, placed anywhere in the excavation zone, can be expected to detect a leak of less than 0.001 gallons (0.0038 liters) per hour over a period of about one month. As shown in Figure 3, the simulated gasoline vapor concentration at even the most distant, shallow sensor was close to the 500 ppm sensor "alarm" level after four weeks.

Increasing the moisture content or decreasing the backfill porosity slowed the simulated detector responses, as depicted in Figure 4 for the deep sensor halfway around the tank from the leak. At this location, however, total gasoline vapors were simulated to exceed 500 PPM by one month, even for the wettest, least porous conditions.

Vapor detectors appear to be potentially effective leak sensors. Comparing the simulation results to detection performance standards yields quantitative justification for using vapor monitors. For

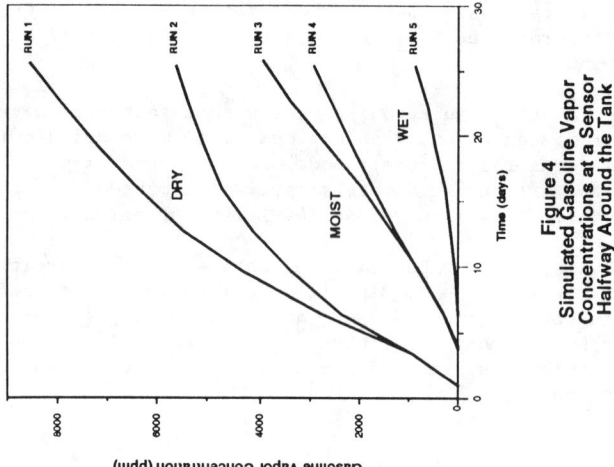

Figure 4
Simulated Gasoline Vapor Concentrations at a Sensor Halfway Around the Tank

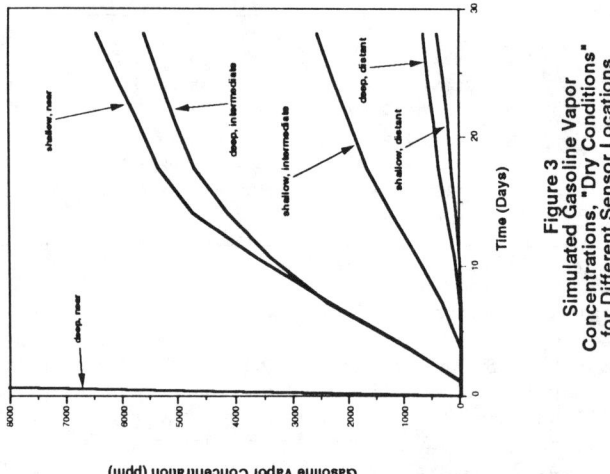

Figure 3
Simulated Gasoline Vapor Concentrations, "Dry Conditions" for Different Sensor Locations

SCHREIBER AND ROSENBERG ON LEAK VAPOR DIFFUSION   81

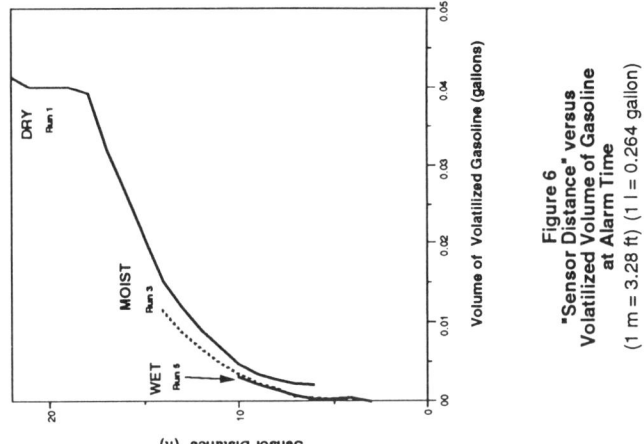

Figure 6
"Sensor Distance" versus
Volatilized Volume of Gasoline
at Alarm Time
(1 m = 3.28 ft) (1 l = 0.264 gallon)

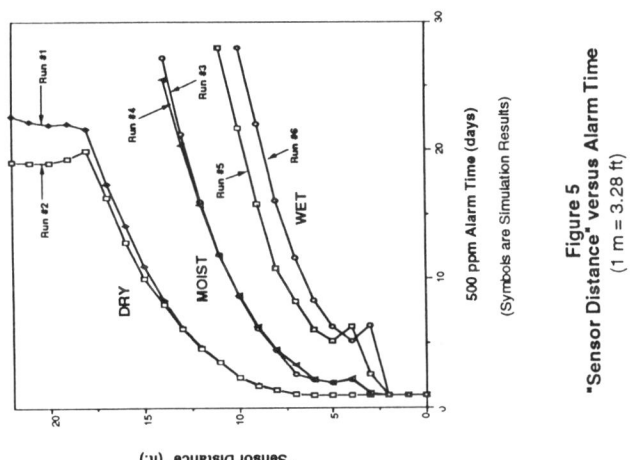

Figure 5
"Sensor Distance" versus Alarm Time
(1 m = 3.28 ft)

82 LEAK DETECTION FOR UNDERGROUND STORAGE TANKS

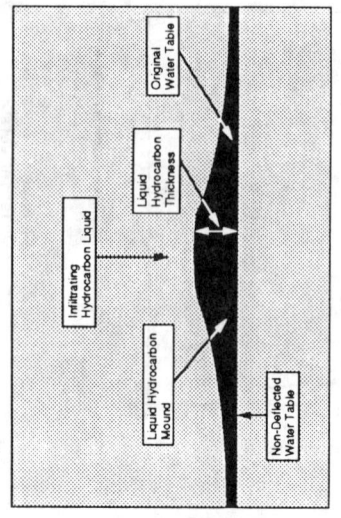

Figure 8
Lens of Liquid Hydrocarbon Upon a Deflected Water Table

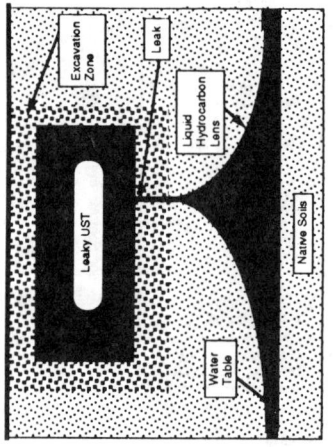

Figure 7
Cross-Section of a Typical Leaky UST Site

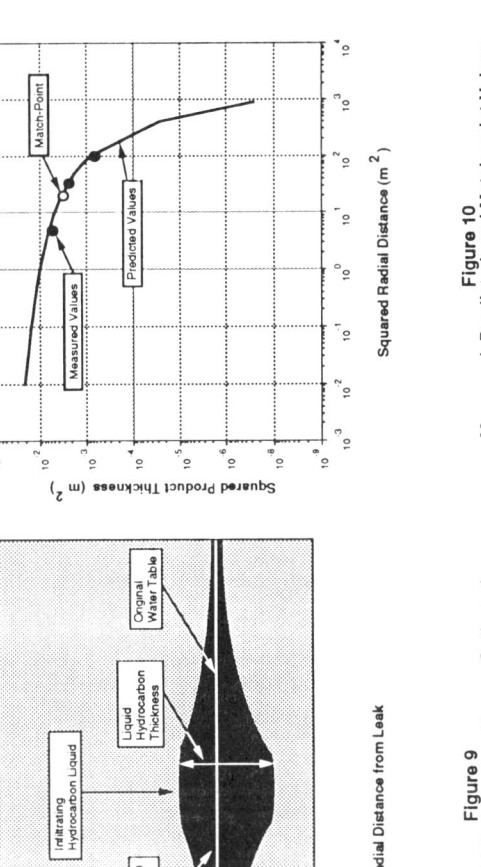

Figure 9
Lens Of Liquid Hydrocarbon Upon a Deflected Water Table

Figure 10
Measured, Predicted, and Match-point Values
(From Parker et al., 1988)

example, with a 30-day testing interval and a detection accuracy of 0.1 gallons (0.38 liters) per hour for in-tank tightness testing, as much as 72 gallons (273 liters) could be released before a leak is detected. On the other hand, if a vapor sensor can detect a leak of about 0.0001 gallons (0.00038 liters) per hour, as simulated by our single-tank vapor diffusion model, less than one-tenth of a gallon would escape undetected.

Analysis of the simulation results also indicates that sensors responding to total hydrocarbons, placed within about 10 feet (3.05 meters) from the leak in terms of "sensor distance," can be expected to respond within a week or so, depending on soil and backfill conditions (Figure 5). Beyond this distance, backfill porosity and moisture content will have a more pronounced effect. This is shown more effectively in an interesting plot of "sensor distance" versus the amount of leaked gasoline when detection occurs (Figure 6). For all simulated conditions, low volumes of leakage trigger alarms up to a "sensor distance" of 10 feet (3.05 meters), beyond which a significantly higher rate of increase in leaked volume is needed to set off alarms. The design of vapor detector networks could therefore be based on expected "zones of detection," a concept analogous to the "zones of influence" of groundwater pumping wells in an aquifer.

Common sense says that the vapor sensors should be placed as close to the leaks as possible. Without knowing in advance where the leak will be, however, the designer should anticipate where leaks are likely to occur, and where they would be most difficult to detect if they were to occur. Leaks on the bottom of the tank are most likely simple because the bottom of the tank always has product; also, leaks near the top could be expected to result in drips rolling along the sides of the tank to the bottom. This indicates that sensors should be placed lower in the excavation zone, and preferably at or around the depth of the tank or excavation zone bottom. This is the most difficult and expensive monitoring depth, but according to the vapor transport simulations, a shallower sensor at the same plan-view location will respond substantially later. Although gravity-induced flow was not simulated, it would have produced an even greater differential response between deep vapor monitors and shallow ones.

VAPOR MODELING CONCLUSIONS

Vapor transport modeling has demonstrated that external vapor sensors in the tank excavation zone can be expected to act as good "early warning" devices. Vapor sensors in a properly designed network are predicted to be capable of detecting new leaks on the order of 0.0001 gallons (0.00038 liters) per hour within days. This promising evaluation of vapor sensors should be tempered by recalling that the modeling results were based on several simplifying assumptions. The modeling results, however, are a good first step in the quantitative assessment of vapor detector performance.

LIQUID BUILD-UP ANALYSIS

The second approach developed as part of the EPA UST work assignment was for analyzing the build-up of liquid hydrocarbons on the water table. Similar to the vapor diffusion simulations, this work assignment was intended to help provide guidance for detection network design, as well as to provide a tool for corrective action decision-making.

This section of the paper summarizes the approach developed to analyze the spreading of fluids which are immiscible in water and are less dense than water. These fluids, such as gasoline and other petroleum fuels, tend to pool upon the water table.

The development and propagation of a mound of liquid hydrocarbon upon a water table below a leaking UST was addressed with an analytical technique that constitutes a pragmatic approach to modeling of liquid hydrocarbon flow in the subsurface. Using this analytical technique, the leak rate, volume, and duration can be estimated based on measurements of liquid hydrocarbon thickness in monitoring wells. A similar analytical technique, for example, was used by Holzer [8] to describe the movement of an oil spill on the water table.

The expression for the growth of a groundwater mound from a circular recharge area can approximate the accumulation of liquid hydrocarbon upon the water table (Figure 7). To do this, the kinematic viscosity of the fluid must be estimated for the liquid hydrocarbon. The hydraulic conductivity of the porous medium is scaled to reflect the permeability of the medium to the liquid hydrocarbon.

General assumptions are also made to solve this problem:

- the aquifer is infinite in areal extent;
- the aquifer is homogeneous and isotropic;
- the leak flows at a steady rate;
- no liquid hydrocarbon is lost to dissolution or volatilization;
- the measured liquid hydrocarbon in monitoring wells can be corrected to accurately reflect the actual thickness of liquid hydrocarbon in the formation;
- the hydraulic conductivity and the storage coefficient are constant;
- the percolation is constant; and
- the leaked liquid hits the water table quickly (i.e., travel time virtually is zero).

The analyst may assume that the water table is flat or that it is deflected by the liquid hydrocarbon, placing upper and lower limits on the estimates of leak rate and duration. A match-point analysis, similar to aquifer pumping test analyses, is used. The thickness of the pooled hydrocarbon for the two cases -- non-deflected and deflected water table -- is shown in schematic cross-section in Figures 8 and 9.

The first case is analogous to the development of a groundwater mound in response to percolating water from a circular recharge basin. This case was analyzed by Hantush (1967)[9], for which he provided a solution based on type-curve fitting. The major assumption is that the water table remains flat below the pooling hydrocarbon. The solution to the second, deflected water table case is taken from Hantush's solution to the freshwater lens problem (Hantush, 1968)[10]. In this analogy, the pooled hydrocarbon is taken to be analogous to freshwater sitting atop denser saltwater. In this case, the fluid conductivity can be corrected, based on the density difference, to yield an "apparent fluid conductivity." This is then used in place of the actual fluid conductivity in the analysis. All other aspects of the analysis are identical to the non-deflected case. These two solutions yield results that bound the likely actual shape of the mound.

The leak rate and duration can be estimated from type-curve fitting by following these steps. First, the hydrocarbon thickness and distance-from-the-UST data are plotted on log-log paper, as shown in the example of Figure 10. These data values were taken from a paper by Parker, et al. (1988)[11]. The type curve is fitted to the data, and a match-point is selected. Using the match-point, the leak rate is obtained from an equation that relates leak rate to the functional value of Hantush's type curve.

The second step is to estimate the duration of the leak. This is done by performing an iterative calculation of leaked volume, based on a numerical integration of the mound shape. This is eased by the use of a log-log plot of hydrocarbon thickness versus product volume, developed from tabulated values of the standard well function. Then, having an estimate of leaked volume, the duration of the leak is estimated by dividing leak rate into leaked volume.

For the example data shown in Figure 10, the techniques described above produced an accurate estimate of the amount of leaked product. The analyses produced an estimate of about 7 800 liters of leaked hydrocarbon liquid, whereas the volume estimated from field measurements of product thickness was about 6 700 liters.

Information on the actual leak rate and duration, unfortunately, was not available, although the technique produced a realistic range of estimated rates and durations. Leak rate was predicted to range from 1.6 to 8.2 liters per hour, and duration from 40 to 200 days. These estimates appear to be consistent with the description of the case given by the authors.

In fact, verification of these techniques was hindered by the lack of available, complete sets of data that contain known values for at least two of the three parameters -- leak rate, leak duration, and measured product thickness or volume. It is recommended, therefore, that complete data sets be sought to test this approach further.

It is also important to note that these techniques can be used to predict the areal extent of liquid hydrocarbon, by running the analyses "backward." In effect, the techniques can provide estimates of one of the three parameters, assuming that data are available to describe the other two. For example, with measurements or estimates of leakage rate and duration, the thickness of the pooled liquid can be estimated, as well as the extent of the free product.

Finally, the limitations of this approach must be realized by those using it. In particular, the results are probably not realistic for sites where the porous media are fine-grained, with strong capillary effects. The techniques are applicable, however, to many UST sites where the soils are relatively permeable, and the water table is below a relatively dry excavation zone.

These techniques have been developed as a practical, simple way of estimating leak rate, duration, and product thickness, and should be used only in performing approximate estimates of these parameters in appropriate hydrogeologic settings. If used properly, they should offer a powerful means of performing initial site assessments of volume of leaked product and its extent, and of "back-calculating" the characteristics of the UST leak source.

OVERALL CONCLUSIONS

The techniques presented above have been used to help guide the development of UST leak detection regulations, and can be used to perform estimates of hydrocarbon vapor concentrations and liquid build-up on the water table. Use of these techniques requires that the analyst be aware of the underlying assumptions, which restrict the applicability of the solutions. Those limitations notwithstanding, the techniques offer a powerful and pragmatic, yet simple means of developing planning-phase estimates of leakage rate, leakage volume, rate of spreading, and depth of product. Such analyses can be a valuable step towards understanding UST-contaminated sites, and building the knowledge needed to remediate and monitor such sites.

DISCLAIMER NOTICE

Although the research described in this article has been supported by the U.S. Environmental Protection Agency, through Contract Number 68-03-3409 to Camp Dresser & McKee/Federal Programs Corporation, this paper has not been subjected to agency review and therefore does not necessarily reflect the views of the agency and no official endorsement should be inferred.

ACKNOWLEDGEMENTS

This work was supported by EPA's Office of Research and Development (EPA/ORD) under a contract entitled "Research for Abatement of Leaks for Underground Storage Tanks Containing Hazardous Substances." Dr. Philip Durgin acted as the Technical Project Monitor.

REFERENCES

[1] Schreiber, R.P., B.S. Levy, and M.S. Rosenberg, "Simulation of Subsurface Vapor Movement from a Leaking Tank," International Conference on Advances in Groundwater Hydrology, American Institute of Hydrology, Tampa, Florida, November 16-18, 1988.

[2] Levy, B.S., P.J. Riordan, and R.P. Schreiber, "Estimation of Leak Rates from Underground Storage Tanks," Journal of Ground Water, v. 28, n.3, May-June 1990, Dublin, Ohio.

[3] Camp Dresser & McKee Inc., June 13, 1988. "Modeling Vapor Phase Movement in Relation to UST Leak Detection," Final Report, Work Assignment #4, Contract No. 68-03-34-09 with U.S. Environmental Protection Agency.

[4] Millington, R.J., and Quirk, J.M., 1961. "Permeability of Porous Solids," Transactions of the Faraday Society, Volume 57, Page 1200.

[5] DYNFLOW A Three-Dimensional Finite-Element Groundwater Flow Model; Description and User's Manual, Version 3.0 (draft), Camp Dresser & McKee Inc., Boston, Massachusetts, November 1984.

[6] Levy, B.S., R.P. Schreiber, and M.S. Rosenberg, "Modeling Vapor Transport for Evaluating Tank Leak Detectors," presented at National Water Well Association Conference on Petroleum Hydrocarbons and Organic Chemicals in Ground Water: Prevention, Detection, and Restoration, Houston, Texas, November 1988.

[7] Schreiber, R.P., B.S. Levy, and M.S. Rosenberg, "Modeling Vapor Transport for Evaluating Tank Leak Detectors," Second Annual Conference on Underground Storage Tank Management and Hydrocarbon Contamination Cleanup, conducted by Resource Education Institute, Sturbridge, Massachusetts, April 1988.

[8] Holzer, T.L., "Application of Ground-Water Flow Theory to a Subsurface Oil Spill," Journal of Ground Water, v. 14, p. 138-145, 1976.

[9] Weber, D.W., and Stetzenbach, K., 1988. "Vapor Transport Model Validation of Flow and Transport Models for the Unsaturated Zone," Ruidoso, New Mexico, May 1988.

REFERENCES (cont.)

[10] Hantush, M.S. 1967. "Growth and Decay of Groundwater-Mounds in Response to Uniform Percolation," Water Resources. v. 3, no. 1, pp. 277-234.

[11] Hantush M.S. 1968. "Unsteady Movement of Fresh Water in Thick Unconfined Aquifers," Bulletin International Association of Scientific Hydrology. v. 13, no. 3, pp. 40-60.

[12] Parker, J.C., J.J. Kaluarachchi, and A.K. Katyal. 1988. "Areal Simulation of Free Product Recovery From a Gasoline Storage Tank Leak Site," Proceedings of Petroleum Hydrocarbons and Organic Chemicals in Groundwater: Prevention, Detection, and Restoration, Houston, TX, Nov. 9-11, 1988. pp. 315-334.

Marc A. Portnoff,[1] Richard Grace,[2] Alberto M. Guzman,[3] and Jeff Hibner[4]

EVALUATION OF METAL OXIDE SEMICONDUCTOR AND POLYMER ADSORPTION GAS SENSORS AS APPLIED TO UNDERGROUND STORAGE TANK LEAK DETECTION

REFERENCE: Portnoff, M. A., Grace, R., Guzman, A. M., and Hibner, J., "Evaluation of Metal Oxide Semiconductor and Polymer Adsorption Gas Sensors as Applied to Underground Storage Tank Leak Detection," Leak Detection for Underground Storage Tanks, ASTM STP 1161, Philip B. Durgin and Thomas M. Young, Eds., American Society for Testing and Materials, Philadelphia, 1993.

ABSTRACT: Gas sensor properties are measured with the purpose of comparing two sensor technologies used for underground storage tank leak detection. Metal Oxide Semiconductor (MOS) and the Polymer Adsorption (PA) gas sensors were tested in simulated underground storage tank environments using the Carnegie Mellon Research Institute (CMRI) automated gas testing facilities. This automated system was used to monitor the response of the sensors while dynamically exposing them to various mixtures of methane, butane and xylene. The sensors were also tested to determine the effects of humidity on their responses. Sensor responses were characterized by sensitivity, selectivity, and speed of response and recovery to the test environments. The test results are presented as a list of sensor specifications to allow the potential end user a direct comparison of these two different types of sensors.

KEYWORDS: metal oxide semiconductor gas sensor, polymer adsorption sensor, vapor sensor testing, underground storage tank

Over two million underground storage tanks (UST) are currently being regulated by the EPA. By 1993 the vast majority of these tanks are to be equipped with one or more of the approved leak detection

---

[1] Manager of the Sensor/Laser Laboratory, Carnegie Mellon Research Institute, 4400 Fifth Avenue, Pittsburgh, PA 15213
[2] Project Manager, Advanced Devices and Materials Group, Carnegie Mellon Research Institute, 4400 Fifth Avenue, Pittsburgh, PA 15213
[3] Director, Advanced Devices and Materials Group, Carnegie Mellon Research Institute, 4400 Fifth Avenue, Pittsburgh, PA 15213
[4] Laboratory Technician, Carnegie Mellon Research Institute, 4400 Fifth Avenue, Pittsburgh, PA 15213

methods. Vapor monitoring equipment, housed in monitoring wells surrounding the UST, is one of the options for protecting the environment from leaking gasoline tanks or fuel spills.

The concept behind vapor monitoring, is that a small leak of a volatile liquid will generate a large increase in product vapor concentration. By proper placement of the monitor wells, taking into account adequate soil porosity, the product vapor will readily migrate to the monitoring wells. There, the increased vapor concentrations will be detected by the vapor sensor and sound an alarm.

In practice, however, this concept has been difficult to implement because of some misunderstandings in the properties of commonly used vapor sensors. These include how sensitive a sensor is to leaking gasoline; how insensitive a sensor is to the background gases found at UST sites; and how to properly calibrate a sensor.

This study was initiated by the EPA Office of Underground Storage Tanks to help the UST community to better understand the capabilities and limitations of commercial vapor sensors used in continuous vapor phase product leak detectors. The intent was to demonstrate that a laboratory method of evaluation could provide this information.

Properties that define a vapor sensor's performance include: 1) sensitivity to a target gas or class of gases; 2) selectivity so that the sensor readings are not misinterpreted; 3) response and recovery to the target gas in a reasonable time period, and 4) stability during the time of data collection. By understanding these properties, the appropriate sensor can be selected for a given environment. However, it should be made clear that the same vapor sensor can work well in one application and poorly in another.

For these series of tests the determination of the UST environment was based upon a study performed by Geoscience Consultants, Ltd., in 1988 [1]. This study detailed the hydrocarbon vapor concentration at 27 gasoline service stations from three diverse geographic regions in the United States. Their findings indicated that; 1) all the surveyed locations had some evidence of underground methane and gasoline vapor products (background vapors); 2) methane existed in high concentrations at many locations; 3) tracking butane concentrations would be useful in detecting recent gasoline leaks or spills, and 4) that m-xylene was a large component of gasoline product. Also, it was determined that the humidity level at UST sites is considered to be near saturation [2].

In this environment, therefore, what sensor properties need to be measured to determine the appropriate selection of a vapor sensor? Clearly a sensor needs to be sensitive to gasoline. To simplify testing and because the composition of gasoline varies, iso-butane and m-xylene were chosen as tags to identify a possible gasoline leak. Iso-butane was selected to represent the C_4 - C_6 class of hydrocarbons and m-xylene as chosen to represent the benzene-toluene-xylene class of compounds. Both classes of compounds are major constituents in gasoline.

The interference gases selected were methane and water vapor. Methane was select because it is found at many UST sites. If a sensor responds to methane but the instrument's user is unaware of this sensitivity, then, this instrument placed in the field could produce false alarms. Also, it is important to know how a sensor might be affected by changes in humidity. For example, if a sensor is more sensitive to gasoline in a damp environment and if it were calibrated with dry gas, this sensor would respond to gasoline levels less than the preset alarm levels leading to false alarms. A sensor less sensitive in damp air, and calibrated with dry gas would lead to a worse situation in which the sensor would not alarm when gasoline levels reached the preset alarm levels.

How fast a suitable sensor responds to UST leaks needs to be taken into context. Leaks at USTs generally occur slowly, and site monitoring is done on time scales of days and not minutes. Therefore, a fast sensor response is not essential. However, recovery time can be important in situations where an accidental spill occurs. In this case, if a sensor takes too long to recover from the spill, the detection of a true leak could be masked.

The reproducibility of sensor properties is essential in maintaining instrument quality control. For example, when a sensor fails and is replaced, if the replacement sensor unknowingly behaves differently, errors in monitoring a site are very likely. By knowing the limitations of reproducibility of various sensor types, steps can be taken to properly calibrate replacement sensors to assure the monitoring equipment's performance.

Two of the most commonly used types of commercial vapor sensors for UST monitoring were selected for this study: the Metal Oxide Semiconductor (MOS) sensor [3] and the Polymer Adsorption (PA) sensor [4]. The properties of these sensors were measured and their performance analyzed with respect to selected test concentrations of methane, butane, and xylene.

EXPERIMENTAL:

The data presented were collected using the CMRI automated gas sensor characterization facility. The facility has been designed to study the behavior of gas sensors and characterize their responses in terms of sensitivity, selectivity, speed of response and recovery, and stability. A computer controlled gas delivery and data acquisition system (GDS) creates the test atmosphere in the sensor test chamber and records the corresponding sensor responses (Figure 1). The GDS controls and sets proper levels of oxygen, nitrogen, and water vapor to create a clean baseline environment through a network of mass flow dilution modules. This clean air can then be contaminated with up to five different vapor compounds. For this study, the facility was modified to independently set concentrations for methane, (CH_4), butane (C_4H_8), and m-Xylene (C_8H_{10}). The GDS was set to maintain a constant flow rate of one liter/minute.

A second gas system, delivering clean humidified air, was used to maintain the sensor atmosphere when the sensor chambers were not connected to the GDS.

An on-line gas chromatograph was used to verify the delivery of gases to the test chamber both during and in between tests.

Test chambers were built to house the sensors. All the materials used in the construction of the chambers were chosen to minimize undesirable out-gassing that might contaminate the test atmosphere. The chambers were also wired to power the sensors and monitor their responses in accordance with manufacturer literature. The volume of each test chamber was 1.2 liters.

Test chamber temperatures were monitored during testing. The PA test chamber temperature operated at room temperature, 22°C ± 1°C. The MOS test chambers' ran hotter, at 33°C ± 1°C, due to the local heating induced by the MOS sensors' operating power requirements.

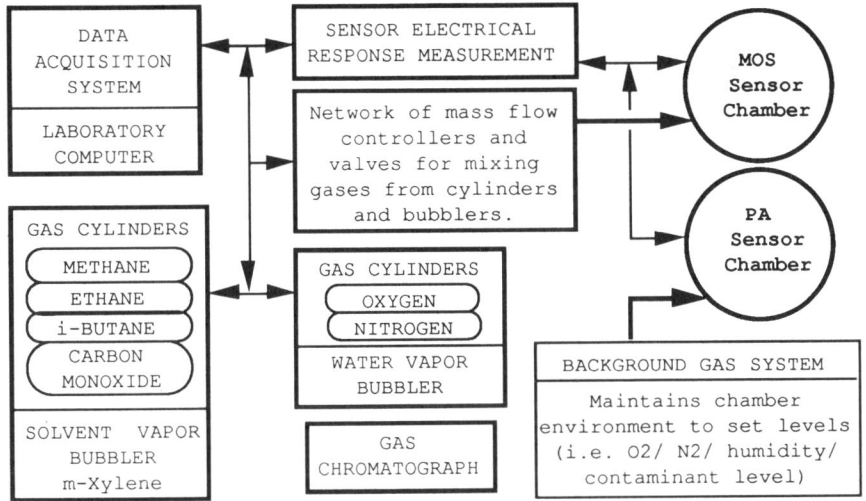

FIG. 1--CMRI Gas Sensor Characterization Facility.

SENSOR CONSTRUCTION AND MODEL EQUATIONS

To simplify direct comparison of these sensors, mathematical models were used to convert non-linear sensor signals into gas concentration (ppm). The model chosen for the PA sensor is the one suggested by the manufacturer [4]. The model selected for the MOS sensors is commonly used in the literature [5].

The Polymer Adsorption Sensor

The PA sensor looks like a small resistor. It is specially coated to make it sensitive to gas vapors. The PA sensor requires no power to operate and is monitored by measuring its resistance like a common resistor.

The base of the coating is a non-conductive, resilient polymer which holds in place conductive particles. The phenomenon of adsorption is the basis for the sensor's sensitivity. In an ambient air environment, the particles, each independently anchored to the polymer surface, are in contact with each other forming an electrical path. When a contaminant vapor comes in contact with the particle surface, a mono-layer is adsorbed onto its surface. Van der Waal's adsorption forces (adhesion of gas molecules to the surface of a solid) cause separation between each of the particles increasing the electrical path's resistance. The electrical resistance measured across a PA sensor is determined by the amount and type of gas molecule adsorbed to its surface [6].

PA sensor data were collected by measuring the sensors' electrical resistance. The resistance is related to concentration for most gas vapor concentrations by equation 1.

$$R = R_b 10^{c/k} \tag{1}$$

where

R = Measured resistance, Ω,
R_b = Resistance in clean air, Ω,
k = Gas constant at ambient temperature, and
c = Gas concentration, ppm.

The PA sensor resistance versus concentration is reported to be a straight line when plotted on a semi-log graph [4].

The model was tested for xylene by exposing the sensors to a xylene ramp of 100 ppm to 1 000 ppm in 100 ppm steps. The resistance versus xylene concentration curve is plotted in Figure 2. This curve is not a straight line. This may indicate that the sensor is not sufficiently sensitive to the lower xylene concentration range.

Because the sensors did not respond to the lower test concentration, a two point fit between the 100 and 1 000 ppm xylene was used to determine R_b and k in equation 1. Solving equation 1 for c yields equation 2 which is used to translate the measured PA resistance into a measured gas concentration.

$$c = k \log 10 (R/R_b) \tag{2}$$

The two point model fit is plotted in Figure 2.

The Metal Oxide Semiconductor Sensor

The MOS sensor is primarily composed of tin oxide sintered on a small ceramic tube. Noble metal wires are used to provide electrical contact between the sintered tin oxide and the electronics used to measure its resistance. The noble metal wires also provide mechanical support. Through the center of the ceramic tube, a coiled wire is positioned to serve as the sensor heater.

The MOS sensors require a small amount of power to operate the sensor element at elevated temperatures, above 200°C. By varying the composition of the sensor element and/or the operating temperature, the sensor's response to various combustible gases can be altered.

The MOS sensors respond to changes in the partial pressure of oxygen. At a set oxygen level, oxygen is adsorbed on the surface of the gas sensing MOS sensor. This adsorption of oxygen on the semiconductor is strong enough to promote electron transport from the semiconductor to the adsorbed oxygen. In the presence of a fixed oxygen environment such as ambient air, an equilibrium state is achieved and the sensor's electrical resistance (baseline) is established. If the environment is then contaminated with a combustible type gas, a surface catalyzed combustion reaction occurs. This reaction causes the surface adsorbed and negatively charged oxygen to be reduced, returning the shared electron to the semiconductor, and decreasing the semiconductor's electrical resistance. The relationship between the change in resistance to the concentration of a combustible gas is non-linear and can be expressed by a power law equation.

For this project, the sensors were powered and measured according to the manufacturer's instructions. Sensor heaters were all powered using a 5 volt power supply. The sensor bias voltage was maintained at 10 volts. Precision load resistors (R_l = 3 920 ohm $\pm 1\%$) were installed in series with the sensor leads. Sensor signals were measured by reading the voltage across the load resistor and converted to sensor resistance using equation 3.

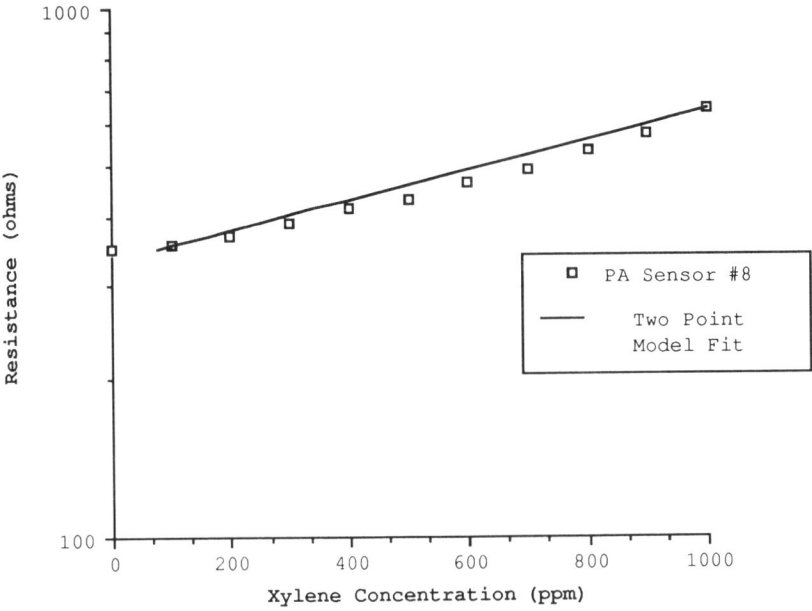

FIG. 2--PA sensor measured and fitted response to xylene
@ 15 K ppm H2O.

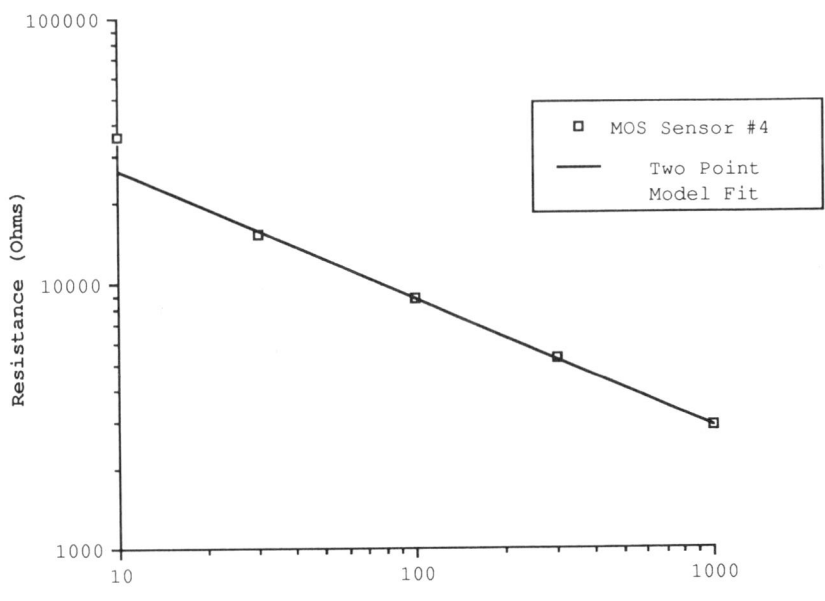

FIG. 3--MOS sensor measured and fitted response to xylene
@ 15 K ppm H2O.

$$R = R_1 (V_B - V_R)/V_R \tag{3}$$

where

R = Resistance, Ω,
R_1 = Load resistor, Ω,
V_B = Voltage bias, V, and
V_R = Voltage across R_1, V.

The resistance versus concentration curve, Figure 3, was observed to be approximately linear on a log - log plot. Therefore, a power law model was adopted for these sensors as seen in equation 4.

$$R/R_o = c^\beta \tag{4}$$

where

R = sensor resistance, Ω,
c = gas concentration, ppm,
β = power law slope, and
R_o = sensor resistance when $c=1$, Ω.

The two parameters R_o and β are determined by considering measurements taken at $c = 100$, and $c = 1\ 000$ ppm for the gas in question. Once the parameters are determined, the sensor resistance is translated into concentration by inverting equation 4 and shown in equation 5.

$$c = \frac{R}{R_o}^{\frac{1}{\beta}} \tag{5}$$

A plot showing how the model fits the sensor response for a MOS is shown in Figure 3.

TEST DESCRIPTIONS

Four specific tests were performed to characterize sensor responses. Each of the following tests were designed to measure one or more specific sensor properties:

Gas Concentration Ramp Test

The Gas Concentration Ramp Test measures a sensor's sensitivity and selectivity to individual test gases. The test exposes the sensors to individual test gases at five different concentrations. For methane and butane the concentrations selected were 50, 150, 500, 1 500, 5 000 ppm. For xylene, because of its lower vapor pressure, the concentrations selected were 10, 30, 100, 300, and 1 000 ppm. Each concentration was held for thirty minutes before proceeding to the next level. The sensors were exposed to clean air for two hours between each ramp.

Ramp tests were performed at two humidity levels. The first set was conducted at 15 000 ppm of water vapor. This level was chosen to represent the humidity present at underground storage sites (97% RH at 55°F). The second set was done in dry air (less than 50 ppm water vapor) to simulate sensor response when exposed to dry calibration gases.

Target Gas Excursion Test

The Target Gas Excursion Test determines how the presence of multiple test gases affects the sensor's sensitivity and selectivity. The test creates a background test atmosphere composed of 500 ppm methane, 500 ppm butane, and 100 ppm xylene in air containing 15 000 ppm of water vapor. During the test, each gas is then individually increased to 10 times its background level for thirty minutes.

Water Vapor Excursion Test

The Water Vapor Excursion Test measures sensor response to the changes in humidity in the presence of multiple test gases. The tests creates the same background test atmosphere used in the Target Gas Excursion Test. The water vapor concentration is then changed in thirty minute steps from 15 000 ppm, to 5 000 ppm, to 1 667 ppm, to 0 ppm water vapor, and then set back to 15 000 ppm.

Response and Recovery Time Test

The Response and Recovery Time Test determines how fast a sensor responds to changes in gas concentration. The tests were performed in air humidified to 15 000 ppm water vapor. The sensors were measured at one minute intervals during the test. The xylene concentration changed in thirty minute steps from 0 ppm, to 1 000 ppm, to 100 ppm, to 1 000 ppm and back to 0 ppm.

The response time is defined as the time required for the sensor to read 95% of its final reading. For this series of tests, the final reading is measured at 30 minutes after the new gas concentration is introduced. The recovery time is defined as the time needed for the sensor to reach 95% of the total change in the sensor reading. For example when changing from 1 000 to 100 ppm xylene, the 95% recovery time would be the time interval at which the sensor reads below 145 ppm xylene.

RESULTS and DISCUSSION

The set of tests, previously described, bracket a range of conditions that vapor sensors are likely to see in the real world. The measured sensor responses are converted from resistance to ppm units using the model equations. Each sensor was fitted individually with a two point calibration.

Tables 1 and 2 summarize PA and MOS sensor performance as a list of sensor specifications. The data are the average of nine PA sensors, and six MOS sensors. The data are reported as the average measured sensor response along with the percent standard deviation.

The following sections offer a brief review of the tabulated data with some graphical examples of how the above data was derived and related to sensor properties.

Reproducibility

The MOS sensors tested showed wide scatter in the sensor model parameters and measured responses. A spread in percent standard deviation ranged from 15% to 100% (Table 1).

The PA sensors tested had model parameters and sensor responses within 11% of each other (Table 1).

98 LEAK DETECTION FOR UNDERGROUND STORAGE TANKS

TABLE 1 -- MOS and PA sensor specifications

Model Parameters @ 15K ppm H2O	MOS Sensor			PA Sensor	
	Avg.	% Dev.		Avg.	% Dev.
B	0.56	21%	K	2988	10%
Ro	9.1E+04	40%	Rb	3.5E+02	10%
Xylene Readings @ 15K ppm H2O *Calibrated at 100 &1000 ppm Xylene					
Xylene Delivered (ppm)	Avg.	% Dev.		Avg.	% Dev.
10	11	54%		61	5%
30	43	23%		68	3%
100*	100	0%		100	0%
300	240	15%		233	2%
1000*	1000	0%		1000	0%
Xylene Readings @ 0 ppm H2O					
Xylene Delivered (ppm)	Avg.	% Dev.		Avg.	% Dev.
10	0	--		119	11%
30	1	100%		126	10%
100	6	75%		139	9%
300	39	55%		251	4%
1000	438	31%		998	1%
Cross Sensitivity (ppm Xylene) @ 15K ppm H2O					
	Avg.	% Dev.		Avg.	% Dev.
5000 ppm Methane	23	37%		63	6%
5000 ppm Butane	793	100%		62	5%
95% Response Time (Minutes) @ 15K ppm H2O					
	Avg.	% Dev.		Avg.	% Dev.
0 to 1000 ppm	15.3	42%		7.3	19%
100 to 1000 ppm	10.2	69%		7.8	20%
95% Recovery Time (Minutes) @ 15K ppm H2O					
	Avg.	% Dev.		Avg.	% Dev.
1000 to 100 ppm	3.3	31%		> 30	0%
1000 to 0 ppm	4.1	23%		> 30	0%

TABLE 2 -- MOS and PA sensor response to Gas and Water Excursion Tests
Calibrated for Xylene @ 15 K ppm H2O

Actual H2O (ppm)	Actual Methane (ppm)	Actual Butane (ppm)	Actual Xylene (ppm)	MOS Sensor		PA Sensor	
				Avg. (ppm)	% Dev.	Avg. (ppm)	% Dev.
15000	500	500	100	299	62%	154	6%
15000	5000	500	100	321	73%	142	6%
15000	500	500	100	308	74%	137	6%
15000	500	5000	100	1043	104%	142	6%
15000	500	500	100	292	75%	135	7%
15000	500	500	1000	1720	40%	941	1%
15000	500	500	100	280	77%	213	6%
15000	500	500	100	273	68%	134	8%
5000	500	500	100	157	65%	137	9%
1667	500	500	100	100	59%	131	11%
0	500	500	100	58	53%	127	11%
15000	500	500	100	318	78%	111	10%

Sensitivity

Gas concentration ramp tests were used to determine the test gas to which the sensors were most sensitive. The sensors were then modeled for this target gas.

The PA sensor's measured responses to xylene, butane, and methane concentration ramps are plotted in Figure 4. The sensor clearly responded to xylene at concentrations over 100 ppm as shown by its increased resistance. The sensor's resistance did not change when exposed to methane and butane at concentrations up to 5 000 ppm. Thus, the PA sensors were modeled and calibrated for xylene, and their responses reported in terms of xylene concentration (ppm), (Table 1).

The PA sensor model does not exactly fit the data indicating that the sensor was not sufficiently sensitive to the lower xylene concentration range. Readings of 61 and 68 ppm xylene in the presence of 10 and 30 ppm xylene, respectively, reveal the baseline or zero reading for these sensors. Also, the reading of 233 ppm xylene in the presence of 300 ppm xylene indicates the model is insufficient to truly characterize this sensor. However, the small spread of 2% among the 9 PA sensors indicates that the sensors are responding similarly.

A MOS sensor is plotted with respect to the same xylene, butane, and methane concentration ramps as shown in Figure 5. For this sensor, the resistance decreased with respect to all the test gases. However, it was most sensitive to xylene as seen by the larger changes in resistance at a given concentration level. The MOS sensors are sensitive enough to measure 10 ppm xylene (Table 1) and were therefore modeled and calibrated for xylene.

Water Response

PA sensor sensitivity to xylene is not affected by the changes in the level of humidity. This is indicated in Figure 3 by the overlapping data points for xylene. These points were taken at the wet (15 000 ppm water vapor) and dry (0 ppm water vapor) conditions and quantified in Table 1.

For the MOS, sensor changes in readings of more than 50% were observed when the humidity varied from the wet to dry conditions. This is shown in Figure 5, a resistance versus concentration plot and again in Figure 6, a concentration versus time plot. The plotted lines in Figure 6 show when the test gases are introduced and to what concentration levels. The sensor response is plotted as in ppm of xylene, both for the dry and wet conditions.

Figure 7 plots the response of a MOS sensor and a PA sensor, computed as ppm xylene, during a water vapor excursion test. For the MOS, sensor changes in reading of more than 50% were observed when the humidity varied from wet to dry conditions. The PA sensors showed little effect due to short term changes in humidity (Table 1).

Selectivity

The selectivity of a sensor relates to how the sensor responds to gases, other than the one it is calibrated for, both individually and in mixtures. If a sensor is perfectly selective, it will respond to only its target gas. If the sensor is not perfectly selective, its cross sensitivity is an indication of how a particular gas could cause a false reading.

The average cross sensitivity response of the MOS sensors to 5 000 ppm methane and 5 000 ppm butane, is 23 ppm and 793 ppm respectively

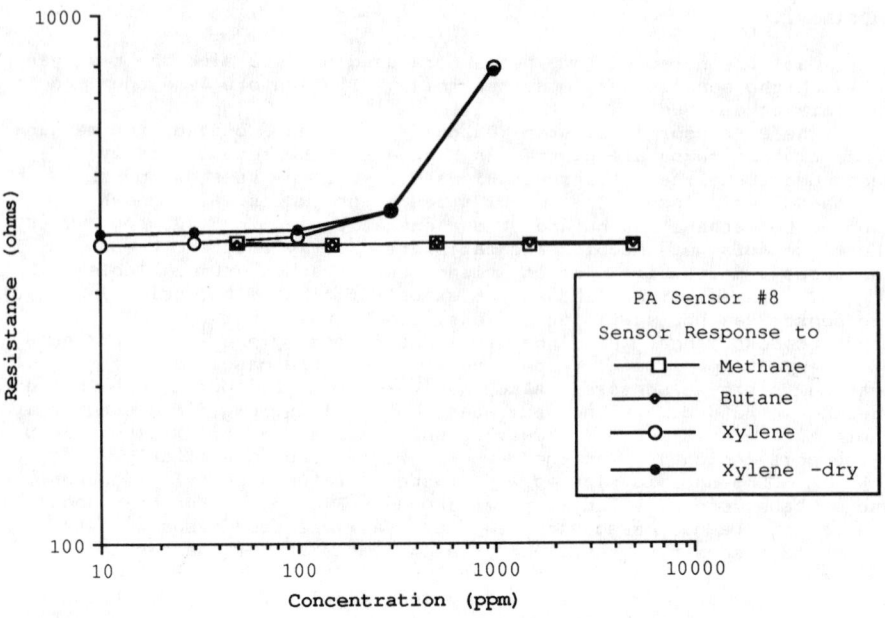

FIG. 4--PA sensor response to methane, butane, and xylene concentration ramps @ 15 K ppm H2O.

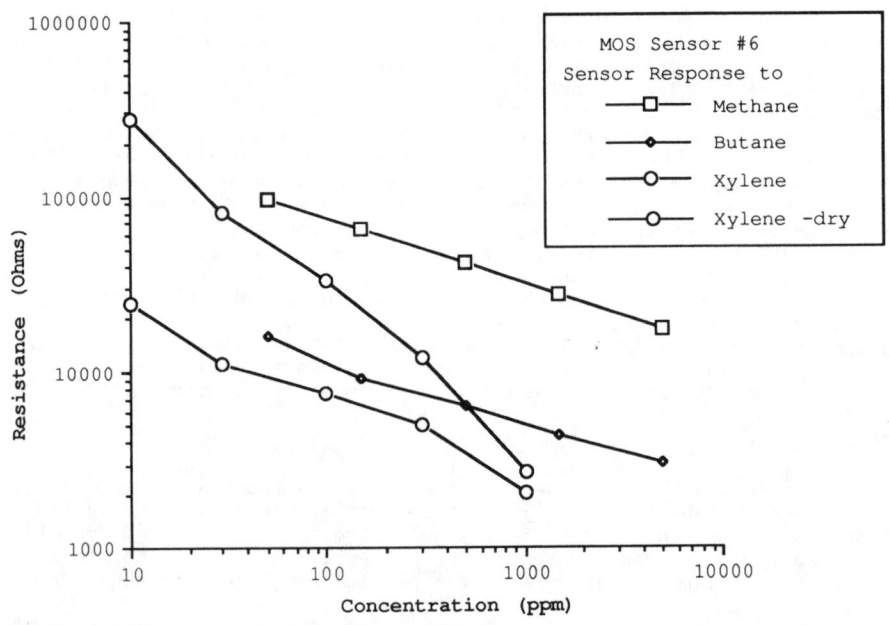

FIG. 5--MOS sensor response to methane, butane, and xylene concentration ramps @ 15 K ppm H2O.

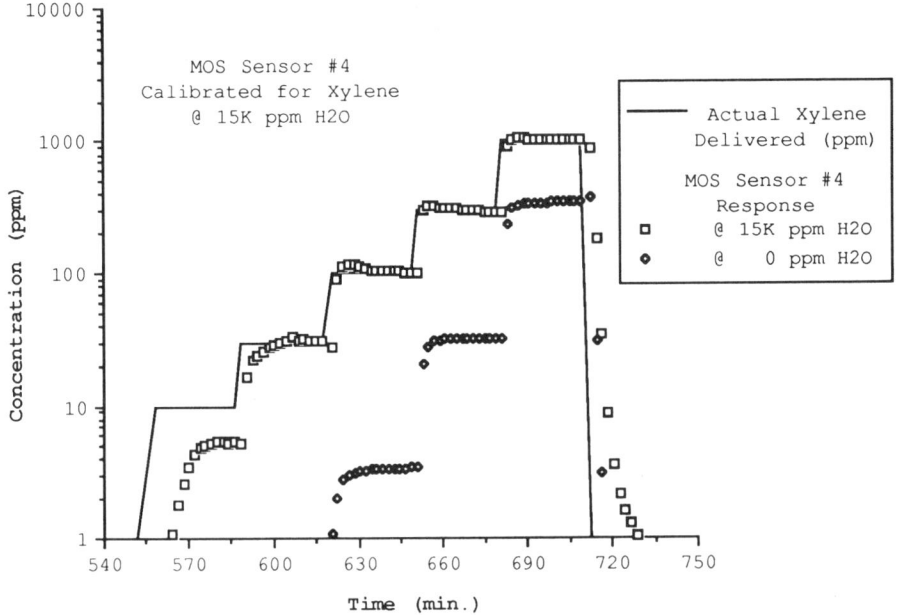

FIG. 6--MOS sensor response to xylene concentration ramps @ 15 K ppm and 0 ppm H2O.

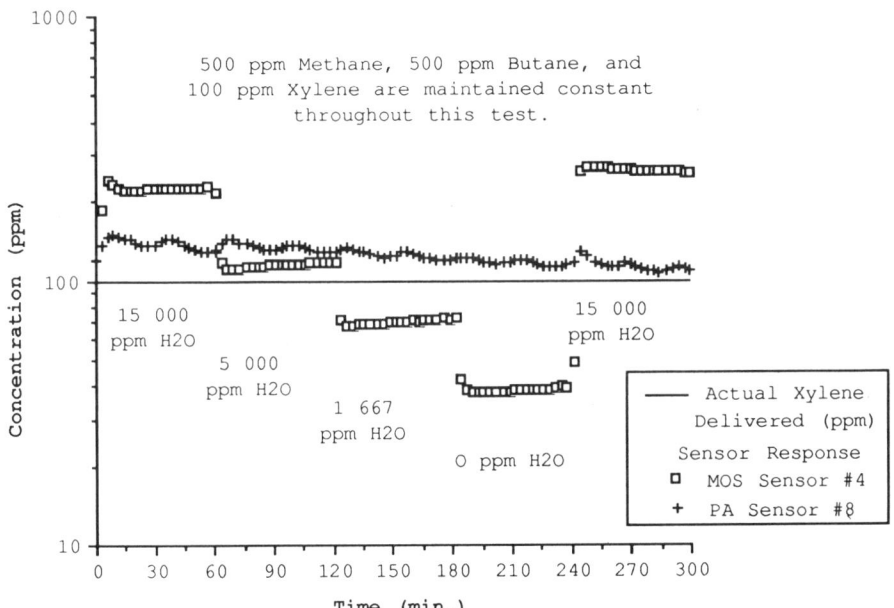

FIG. 7--MOS and PA sensor response to changes in humidity in a mixtures of methane, butane, and xylene.

(Table 1). This indicates a very small cross sensitivity to methane and that the sensor is about six times less sensitive to butane than xylene.

The PA sensor's cross sensitivity response to 5 000 ppm methane and 5 000 ppm butane, is 63 ppm and 62 ppm respectively (Table 1). As mentioned previously, these values indicate a zero response showing the PA sensors to be insensitive to both methane and butane.

The sensor cross sensitivity in multiple gases for the PA and MOS sensors are tabulated in Table 2. The PA sensors are selective to xylene even in the presence of a mixture of methane and butane. This is apparent in that the PA sensor's xylene response does not vary even when the concentrations of methane and butane are increased to 5 000 ppm.

The MOS sensor cross sensitivity to butane is larger in a mixture than would be expected from the tests performed with individual gas ramps. At the background level (500 ppm methane, 500 ppm butane, 100 ppm xylene), the MOS sensor reads over 300 ppm xylene. This error can be attributed mainly to the presence of the 500 ppm butane. The MOS sensor is shown to be insensitive to methane by the slight increase in the xylene level as the methane is increased to 5 000 ppm. When the butane level is raised to 5 000 ppm, the xylene reading increases to over 1 000 ppm, and when the xylene level is raised to 1 000 ppm, the xylene reading is increased to 1 700 ppm.

Figure 8 displays these results showing the response of a MOS sensor and a PA sensor, computed as ppm xylene, during an excursion test.

Speed of Response and Recovery

The MOS sensor's 95% response times are higher when changing from 0 ppm to 1 000 ppm xylene (15 minutes) than from when changing from 100 ppm to 1 000 ppm (10 minutes). The recovery times from either 1,000 ppm to 0 ppm or 1 000 to 100 ppm are about the same at 4 and 3 minutes, respectively.

The PA sensor's 95% response times for the above tests were similar at 7.3 and 7.8 minutes, respectively. The recovery time for the PA sensor was over 30 minutes.

CONCLUSIONS

Sensor specifications directly comparing the two different sensor types, the MOS sensor and the PA sensor, have been presented. This data was generated by testing the sensors in the laboratory using a set of test conditions bracketing the UST environment. The actual simulated test conditions can be broadened, to include gasoline product vapor testing, to provide distinct information to the end user. Also, this type of lab testing can be utilized to include sensor systems before resorting to costly field tests.

The test results show that both sensor types appear to have sufficient properties to be used for UST leak detection. Both respond well to xylene, with the MOS sensor being more sensitive to lower levels than the PA sensor. Both sensor types are relatively insensitive to methane, which is the primary interfering compound underground. The observed butane response for the MOS sensor is not a serious problem since butane is also a component of gasoline. The PA sensors as a group were more reproducible and had a much smaller humidity interference in comparison to the MOS sensors. These two properties make the PA easier to deal with from a instrumentation and calibration point of view.

However, the PA sensors were observed to have longer xylene recovery times than the MOS sensor.

Stability is a major sensor specification not yet studied. It plays an important role in determining how a sensor is employed in UST monitoring. If a sensor changes with time, independent of the actual conditions, it could lead to false alarms and/or not being able to detect a leak. It is recommended that a stability test be undertaken to determine the calibration periods of the sensors and how their characteristics change with time.

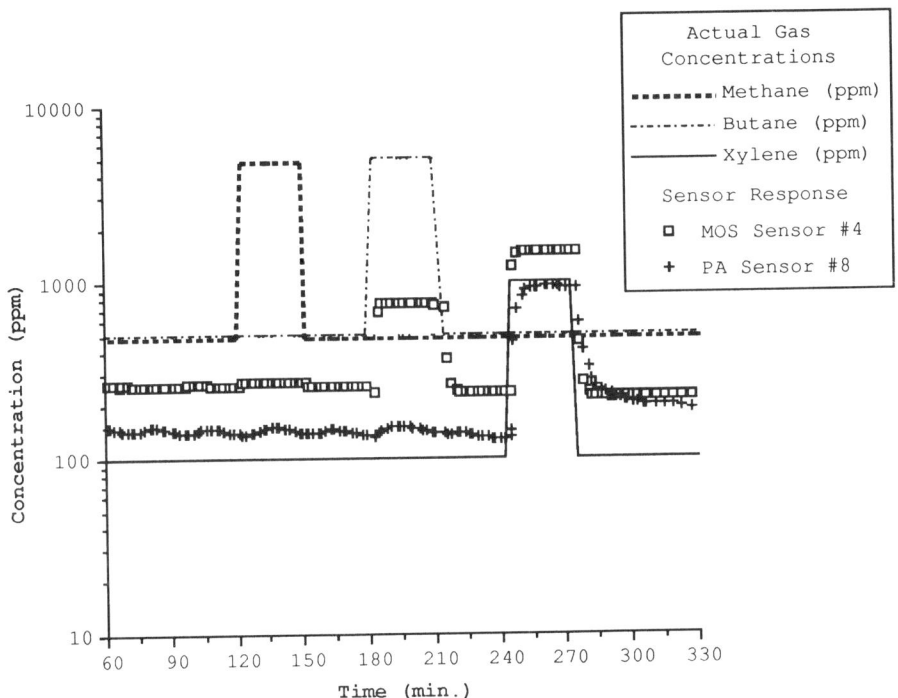

FIG. 8--MOS and PA sensor response to varying mixtures of methane, butane, and xylene @ 15 K ppm H2O.

ACKNOWLEDGEMENTS

This research was funded by the U. S. Environmental Protection Agency, Environmental Monitoring Systems Laboratory, Office of Underground Storage Tanks, Las Vegas, Nevada.

REFERENCES

[1] Schlez, C., "Background Hydrocarbon Vapor Concentration Study for Underground Fuel Storage Tanks," Draft Final Report for U.S. EPA, Contract No. 68-03-3409, February 29, 1988.

[2] Personal communication with Philip B. Durgin, PhD, U. S. Environmental Protection Agency, Environmental Monitoring Systems Laboratory, Las Vegas, Nevada, November 1990.

[3] Figaro Taguchi Sensors are a product of Figaro Engineering of Japan represented by Figaro USA, Inc., P. O. Box 357, Wilmette, IL 60091.

[4] Adsistor Vapor Sensors are products of Adsistor Technology, P. O. Box 51160, Seattle, Washington 98115.

[5] Grace, R., Guzman, A., Portnoff, M., Runco, P., Yannopoulos, L., "Computational Enhancement of MOS Gas Sensor Selectivity," P-33, Proceedings of the Third International Meeting on Chemical Sensors, Cleveland, OH, September, 1990.

[6] Dolan, J., Jordan, W., "Detection Device", U. S. Patent # 3,045,198, July 17, 1962.

Alan E. Grey[1] and Judy K. Partin[2]

FIBER OPTIC CHEMICAL SENSORS - AN OVERVEIW

REFERENCE: Grey A. E. and Partin, J. K., "Fiber Optic Chemical Sensors--An Overview," <u>Leak Detection for Underground Storage Tanks</u>, <u>ASTM STP 1161</u>, Philip B. Durgin and Thomas M. Young, Eds., American Society for Testing and Materials, Philadelphia, 1993.

ABSTRACT: In a span of approximately 20 years, fiber optic sensing has grown from a laboratory oddity to a viable analytical field technique for the detection and monitoring of a wide variety of analytes. One of the reasons for this rapid growth is the range of techniques that can be used for the detection of species. These include changes in absorption, reflection, refraction, phase, polarization, and fluorescence. In general, any chemical or physical reaction that will perturb the light transmission through the optical fiber can be used as the basis for a fiber optic detector. Examples of fiber optic chemical sensors are presented and their advantages over conventional devices are discussed.

KEYWORDS: fiber optic sensors, chemical sensing and monitoring, leak detection, site evaluation, site remediation

The monitoring of underground storage tanks for leaks, and the evaluation and remediation of sites where leakage has occurred, require the development of accurate, inexpensive, in-situ chemical analyses. Rapid on-site detection and analysis minimizes personnel exposure and reduces further transport of pollutants into the environment. It also allows quick feedback during the

[1]Scientific Specialist, Science and Technology Department, Idaho National Engineering Laboratory, Idaho Falls, Idaho 83415

[2]Scientific Specialist, Science and Technology Department, Idaho National Engineering Laboratory, Idaho Falls, Idaho 83415

execution of costly drilling and excavation operations used to locate and remediate contaminated areas.

Fiber optic sensors are uniquely qualified for these applications. They are able to monitor for a variety of analytes incorporating an assortment of detection techniques and geometries. The measurements can be performed with high sensitivity and immunity to electromagnetic interferences. Also, since no electricity is needed at the location of the sensor, they can be used safely in combustible and explosive environments. The ability of fibers to transmit light signals over long distances with low loss allows for the remote placement of sensitive instrumentation away from hazardous or adverse environments. Compact, telemetry-compatible, devices may be engineered from the technology. Fiber sensors are corrosion-resistant and are less sensitive to surface contamination than potentiometric devices since they respond to actual concentrations rather than concentration gradients. Finally, the number of optical parameters including intensity, wavelength, polarization, phase and temporal characteristics that can be varied increases the analytical measurement flexibility relative to devices that rely on electrical transmission. Using these parameters, it is possible to multiplex fiber optic devices into complex arrays, allowing numerous measurements to be performed using the same instrumentation thereby, reducing costs.

An optical fiber, shown in Figure 1, consists of the core, and two coatings, the cladding and the jacket. The purpose of the cladding is to assist in light transmission through the fiber. This is accomplished by using a cladding material with a lower index of refraction than the core material. When light impinges on the cladding it is then reflected back into the core. If the refractive index of

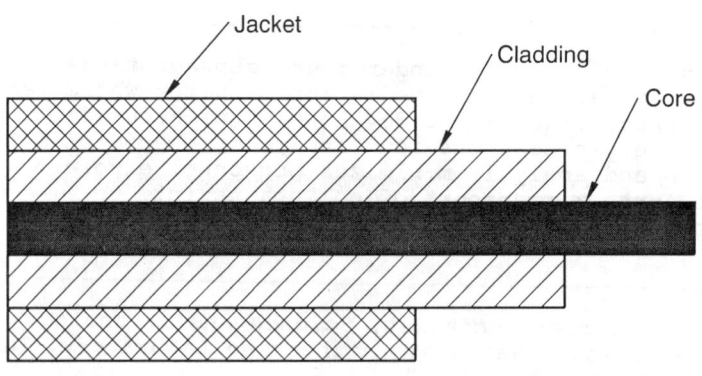

FIGURE 1--Optical fiber

of the cladding is greater than that of the core, the light leaks through the cladding and is lost. Light modulation through the modification of the cladding index is a common technique used in fiber optic sensing.

Commercial optical fibers can be fabricated from either glass or plastic materials. Fused silica is a commonly used core material because of its superior transmission properties and its resistance to chemical attack. The cladding material may be a doped silica or a polymeric material. In chemical sensor work, commercial fibers with polymeric claddings are often chosen because the cladding can then be easily removed and replaced with a designer cladding for the application at hand. The jacket material is typically a tough polymer, although metallic jackets are available for use in high temperature environments. The function of the jacket is to add strength and to protect the fiber during handling. The fibers used in chemical sensors typically vary from 100 to 1000 microns in diameter. The selection of fiber size is dependent upon such factors as numerical aperture, light attenuation as a function of wavelength over the proposed length of the fiber, and the required flexibility of the fiber.

SENSOR TYPES

Fiber optic sensors may be classified into either intrinsic or extrinsic devices, depending upon the configuration of the sensing element. In an intrinsic sensor, the optical fiber itself acts as the sensor. This is often accomplished by modifying the chemistry of the core glass (e.g. by doping the core with a temperature sensitive fluorophore) or by replacing the cladding with specialized materials, such as electro- or magnetostrictive jackets, biological receptors, selective chemical reagent-doped polymers, or reactive metal coatings which enhance the modulation of optical radiation when exposed to specific measurement phenomena. These types of sensors are commonly used to measure such physical phenomena as temperature, pressure, or magnetic and electric field strengths using interferometric detection to compare changes in light phase between a sensing and reference fiber.

In the case of chemical analytes, intrinsic sensing is accomplished by monitoring intensity versus wavelength changes that occur through the interaction of evanescent waves with specialized coatings. Evanescent waves, also called standing or tunneling waves, originate from totally reflected bound waves, as shown in Figure 2. As each bound wave is reflected at the core/cladding interface, a portion of that ray penetrates into the cladding. The depth of pentration is controlled by several factors, but is generally on the order of one wavelength. Thus, the evanescent wave produced in a fiber can be used to interrogate the chemistry applied to the surface of a core or cladding [1-3].

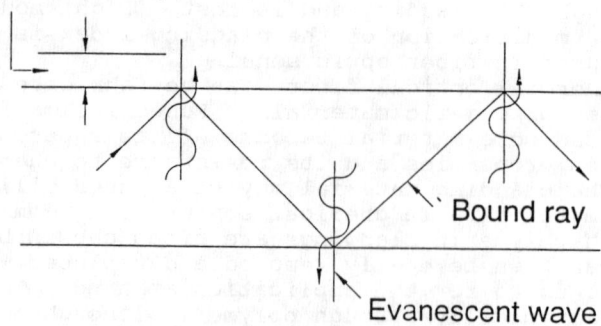

FIGURE 2--Evanescent or standing wave

Extrinsic sensors employ the fiber as a means of transmitting light between the measurement instrumentation and an external transducer, or sampling area, located at the distal end of the fiber. In either case, the sensing materials or reagents may be localized at the sensing region by direct deposition on the fiber or by encapsulation with a polymeric membrane.

The trade-offs between the extrinsic and intrinsic configurations is as follows: In the extrinsic case, a limited amount or area or bound reagent may be sensed; however, the excitation/transmission of the signal to and from the sensing region is efficiently coupled into the fiber. In the case of the intrinsic sensor, a larger amount of reagent is available for sensing, but the coupling via the evanescent wave at the fiber surface is less efficient.

SENSING MECHANISMS

Measurements are typically performed by monitoring intensity variations in a probe beam generated by some attenuation mechanism such as absorption, scattering, or reflection either at the distal tip of a fiber or at the fiber surface, or alternatively, by monitoring fluoresence.

Absorption--When a beam of light impinges upon a material certain wavelengths of the light can be selectively removed by the process of absorption. In this process, photons of appropriate energy interact with an atom or molecule and energy is transferred, producing transitions in their electronic, vibrational and/or rotational energy states. Absorption occurs when the difference in energy states involved exactly matches the energy of the exciting photons.

Many absorption sensors use evanescent wave interactions in an effort to increase detection sensitivity by performing a line-interaction along the fiber length. An example of this type of sensor has been produced at our laboratory for the detection of hydrazine. For this sensor, an approximate 1 meter length of 200 micron quartz fiber was coated with a polymer containing a reactive aldehyde. The polymer chosen had a refractive index slightly less than the 1.46 of the fiber core. The objective was to maintain maximum penetration of the evanescent wave while still maintaining total internal reflectance. When the coated fiber is exposed to hydrazine, a change from colorless to brown is observed. The sensitivity of the sensor can be adjusted by either increasing or decreasing the length of the coated fiber which effectively increases or decreases the amount of chemical reagent available to react with the hydrazine.

A major problem with evanescent wave sensors is the limited penetration of the evanescent wave. Different approaches have been developed to minimize the problem. One such approach is the "sandwich" fiber. A polymer containing the reactive chemistry is applied to the fiber core to form an inner cladding. This polymer is selected with a refractive index greater than or equal to the core. A second porous polymer, with a refractive index less than the core, is then applied as an outer cladding. The porosity allows the analyte of interest to penetrate the outer cladding and interact with the reagent.

Other promising techniques to enhance evanescent field sensing include the use of small (10 micron core), single mode fibers which transmit a significant amount of their optical power outside the core and the use of novel core geometries to couple more energy to the evanescent modes. Another approach to enhancing reagent interaction is through the use of porous fibers. The porous structure is chemically leached into a specially prepared fiber preform generating a large surface-to-area-volume as compared to other fibers. Using cobalt chloride as a color reagent, a humidity sensor has been fabricated and tested based on this technology [4].

The recent availability of fluoride and chalcogenide fibers which transmit near to mid-infrared (2-11 micron) radiation enables the remote sensing of several gaseous hydrocarbon species. These species have a number of prominent vibrational bands, corresponding to hybrid carbon-hydrogen bonds, that occur in the 3-4 micron region. These fibers may be used to transmit infrared radiation to and from sampling cells and standard FTIR spectrometers, or may be used directly through the absorption of the evanescent field, using unclad or permeable clad fiber [5-7].

In reference [8], an optical liquid level sensor for fuel tanks, based on differential optical absorption, is described. The sensor uses a two wavelength radiometric approach, with one wavelength chosen to be strongly absorbed by the liquid and the other weakly absorbed, to cancel out errors which might arise from variations in fuel characteristics, debris build-up on optical surfaces and liquid surface tilts and vibrations.

Refraction/Reflection--At the boundary between two media, light which is not transmitted or absorbed may be reflected or refracted. Modifications in the boundary media which alter its refractance or reflectance properties provide the basis for a number of chemical sensors.

One method of preparing a refractive index based sensor is to remove 2 to 3 centimeters of the cladding and jacket material at the distal tip of the optical fiber. A new polymer is then applied which may be either refractive index matched to the fiber core, or alternatively, to the analyte of interest. In either case, the polymer must selectively absorb the analyte of interest. As the analyte is absorbed into the polymer coating, modifying its refractive index, a change in light is intensity is observed.

Two methods have been used for monitoring this intensity variation. One method places a mirror at the fiber tip to redirect the illumination to the sensing electronics. The other method involves the the attachment of a fluorescent bead at the fiber tip to function as a constant light source. Changes in fluorescence intensity are then observed. A generalized schematic of a fiber optic sensor based on refractive index change is shown in Figure 3.

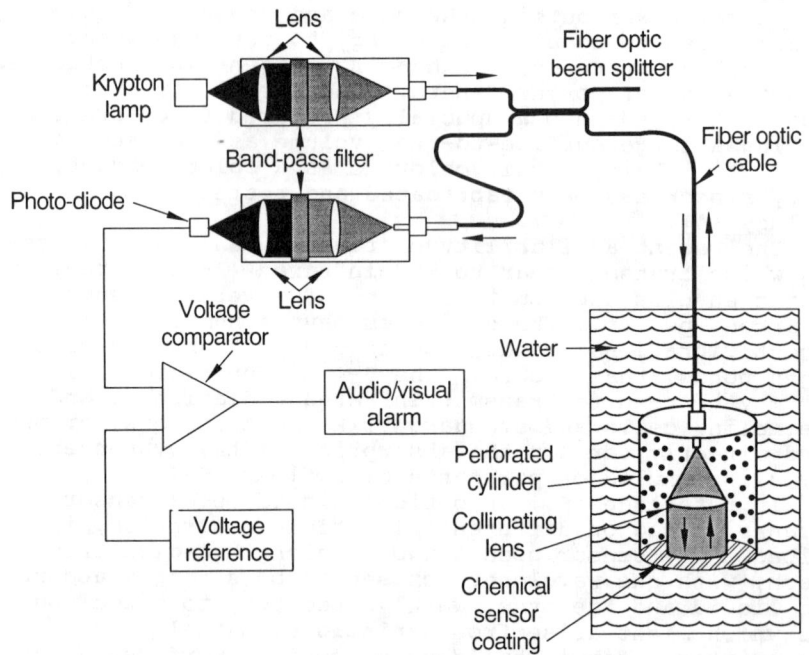

FIGURE 3--Refractive index sensor schematic

A sensor based on this technique has been developed for the detection of light aromatic hydrocarbons by FCI Corporation [9]. It gives a rapid response down to the ppb range and is reversible.

A fiber optic micromirror sensor currently under development consists of a fiber whose tip has been coated with a partially transparent metallic film [10]. On top of this film a thin membrane is deposited. The metallic film forms an interferometric cavity with the membrane which is coupled to the optical fiber. The reflectivity of the metallized tip depends upon the thickness of the membrane layer. Exposure of this layer to various volatile organic compounds results in the swelling of the membrane changing the reflectivity of the fiber tip. This concept provides a generic means for detecting a number of volatile organic compounds including trichloroethylene, methylene chloride, hexane, p-xylene, acetone and carbon tetrachloride.

<u>Fluoresence</u>--Fluoresence is produced when the absorption of energy from a photon causes atoms and molecules to be promoted to a higher energy state. The excited species, however, is short-lived and may release its energy in several ways. One way is through the emission of radiation at a lower frequency. If this emission occurs in a relatively short time frame (e.g. within 100 nanoseconds), it is called fluoresence

Fluoresence-based measurements have several advantages over other techniques. They are very species selective since both the transmitted and the detected wavelengths are characteristic of the constituent of interest. They are very sensitive because of the high cross sections for fluoresence as compared to other scattering processes. The ability to remove the detected wavelength from the excited wavelength makes it easier to discriminate the signal from scattering from constituents.

Laser-induced fluoresence has been used to detect aromatic ground water contaminants [11]. The measurements reported were performed in-situ using optical fibers to transmit the excitation to and fluoresence from the sampling area.

In another type of sensor under development at the Lawrence Livermore National Laboratory, organochlorides are detected using the Fujiwara reaction [12]. In this device, an optical fiber is sealed in a capillary tube, A 10 M potassium hydroxide (KOH) solution is placed around the exposed fiber, followed by a layer of pyridine. The capillary is sealed with a mylar membrane, which allows the passage of organochlorides into the reaction mixture, but inhibits the reactants from leaving the capillary. When the organochloride comes in contact with the pyridine solution a chromophore is produced which can be monitored to indicate the presence of these contaminants.

Fluorescence quenching can also be used as a sensing mechanism. This technique has been used in our laboratory to develop a sensor for trinitrotoluene (TNT). In this sensor, a fluorescent polyaromatic hydrocarbon (PAH) is adfixed to the distal end of a fiber. As the TNT comes in contact with the PAH, the two combine by pi bonding, and the fluorescence from the PAH is inhibited.

As with absorption sensors, fluoresence materials may be bound to the surface of the optical fiber and excited by the evanescent wave [13].

TIME-BASED SENSING

Optical Time Domain Reflectrometry (OTDR) has not been widely applied in fiber optic chemical sensors. However, for certain applications it has enormous potential. Briefly, an OTDR can sense a chemical or physical reaction at the fiber surface, and by timing the return signal indicate the point on the fiber where the reaction occurred [14]. For example, a properly prepared optical fiber installed along a pipeline could monitor for leaks, and indicate where along the pipeline the leak was occurring.

At the present time, OTDR technology is most frequently used for characterization of an optical fiber's uniformity and length, light loss at a splice, location of breaks, and signal attenuation. A schematic of an OTDR is shown in Figure 4.

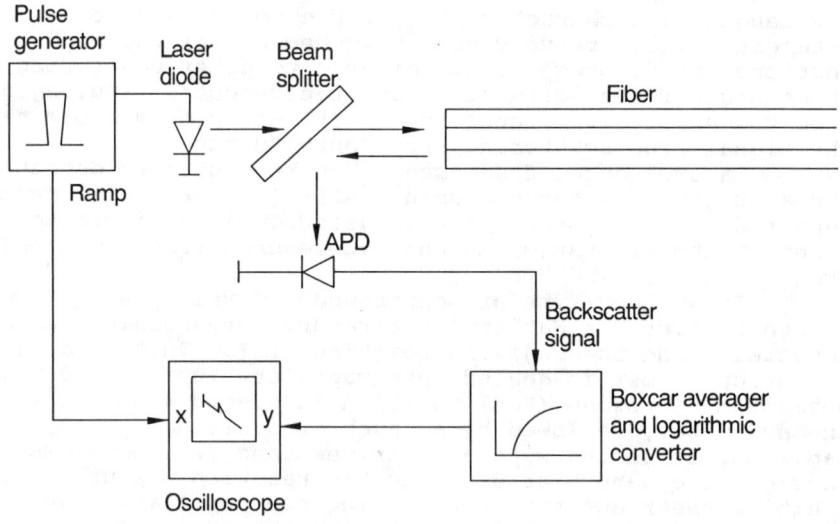

FIGURE 4--Schematic of an OTDR

A pulse generator is used to drive the laser diode. This system sends optical pulses of 10 milliwatts or more into the fiber. The pulse width can range from microseconds to nanoseconds. The repetition rate is usually about 1 KHz when using long fibers, and up to 20kHz when shorter fibers are used. An avalanche photodiode (APD) is frequently used as the detector. The signal from the APD is fed to an amplifier and a digitizer, and then displayed in logarithmic form on an oscilloscope. Two types of backscatter signal can be analyzed, either the Fresnel or the Rayleigh backscatter. The Fresnel reflection is frequently used since it is usually the strongest signal. Its amplitude can be up to 4% of the launch signal.

CONCLUSIONS

Fiber optic chemical sensors represent a versatile and fast growing field of chemical instrumentation. They may be used to detect and monitor for leaks through continuous monitoring of tank liquid level, through identifying gaseous species in interstitial linings and soil samples and by detecting contaminants in groundwater.

They have numerous advantages including: the number of mechanisms which can be used for sensing, the inherent safety of the devices in explosive environments, the ability to perform point or distributed measurements using OTDR techniques, and the large number of analytes which may be sensed.

ACKNOWLEDGEMENT

Work described in this paper was supported by the Department of Energy under Contract No. DE-AC07-76ID0157.

REFERENCES

[1] Wlodarczyk, M. T., Vickers, D. J., and Kozaitis, S. "Evanescent Field Spectroscopy with Optical Fibers Sensing," SPIE, Vol.718, Conference on Fiber Optic and Laser Sensors IV, Cambridge, MA, 1986, pp. 192-196.

[2] Paul, P. H. and Kychakoff, G., "Fiber-Optic Evanescent Field Absorption Sensor," App. Phys. Lett., Vol. 51, 1987.

[3] Sepaniak, M. J., Tromberg, B. J. and Vo-Dinh, T., "Fiber Optic Affinity Sensors in Chemical Analysis," Progress in Analytical Spectroscopy, Vol. 11, 1988, pp. 481-509.

[4] Zhou, Q., Shahriari, M. R. and Sigel Jr., G. H., "The Effects of Temperature on the Response of a Porous Fiber Humidity Sensor," SPIE, Vol. 990, Conference on

Chemical, Biochemical, and Environmental Applications of Fibers, Boston, MA, 1988, pp.153-7.

[5] Saito, M. and Takizawa, M., "Optical Remote Sensing for Hydrocarbon Gases using Infrared Fibers," <u>J. Appl. Phys.</u>, Vol. 63, 1988, pp. 269-272.

[6] Matson, B. S. and Griffin, J. W., "Infrared Fibers for the Remote Detection of Hydrocarbons Operating in the 3.3 To 3.6 Micron Region," <u>SPIE</u>, Vol. 1172, Conference on Chemical, Biological, and Environmental Fiber Sensors, Boston, MA, 1989, pp.13-37.

[7] Ruddy, V., Maccraith, B. D., and McCabe, S., "Remote Flammable Gas Sensing Using a Fluoride Fibre Evanescent Probe," <u>SPIE</u>, Vol. 1267, Conference on Fiber Optic Sensors IV, The Hague, Netherlands, 1990, pp. 97-101.

[8] Yakymyshyn, C. P. and Pollock, C. R., "Differential Absorption Fiber-Optical Liquid Level Sensor," <u>J. of Lightwave Tech.</u>, Vol. 5, 1987, pp.941-946.

[9] Dandge, D. K., et. al., "Fiber Optic Chemical Sensor for Jet Fuel," <u>SPIE</u>, Vol. 1172, Conference on Chemical, Biochemical, and Environmental Sensors, Boston, MA, 1989, pp.132-139.

[10] Butler, M. A., Ricco, A. J. and Buss, R., "Fiber Optic Micromirror Sensor for Volatile Organic Compounds," <u>J. Electrochem. Soc.</u>, Vol. 137, 1990, pp. 1325-26.

[11] Chudyk, W., Pohlig, K., Wolf, L. and Fordiani, R., "Field Determination of Ground Water using Laser Fluorescence," <u>SPIE</u>, Vol. 1172, Conference on Chemical, Biochemical, and Environmental Fiber Sensors, Boston, MA, 1989, pp. 123-9.

[12] Angel, S. M., Daley, P. F. and Kulp, T. J., "Optical Chemical Sensors for Environmental Monitoring," <u>Proc. of Electrochem. Soc.</u>, Vol. 87-9, 1990, pp. 484-489.

[13] Lieberman, R. A., Blyler, L. L. and Cohen, L. G., "A Distributed Fiber Optic Sensor Based on Cladding Fluorescence," <u>J. of Lightwave Tech.</u>, Vol. 8, 1990, pp.212-220.

[14] Kvasnik, F. and McGrath, A. D., "Distributed Chemical Sensing Utilizing Evanescent Wave Interactions," <u>SPIE</u>, Vol. 1172, Conference on Chemical, Biochemical and Environmental Sensors, Boston, MA, 1989, pp. 75-82.

Philip B. Durgin[1]
Ronald W. Michelson[2]

FIELD RESULTS OF HYDROCARBON VAPOR MONITORING TO DETECT LEAKING TANKS

REFERENCE: Durgin, P. B. and Michelson, R. W., "Field Results of Hydrocarbon Vapor Monitoring to Detect Leaking Tanks," Leak Detection for Underground Storage Tanks, ASTM STP 1161, Philip B. Durgin and Thomas M. Young, Eds., American Society for Testing and Materials, Philadelphia, 1993.

ABSTRACT: Vapor monitoring for leak detection around underground storage tanks (UST) is generally regarded as being plagued by false alarms. Examples of field data from six sites are provided. Vapor monitoring indicates real leak or spill events but methods are needed to differentiate between the two. Site assessment, appropriate installation of monitoring points, overfill/spill protection, and data analysis are all important factors for a well-operating vapor monitoring system. Vapor monitoring may gain importance over time as systems are improved, subsurface contamination is reduced, and demands increase for providing assurance that sites are clean.

KEYWORDS: Underground storage tanks, hydrocarbon, vapor monitoring, leak detection, gasoline

Vapor monitoring is the leak detection alternative with the fewest restrictions in the 1988 Federal Regulations for Underground Storage Tanks. The reason for its favored status is that it has the greatest potential to rapidly detect leaks -- particularly small ones. Nevertheless, vapor monitoring is in disfavor nationwide by owner/operators and even many vendors. Their apparent objection is the frequent occurrence of false alarms. On the other hand, many state regulators still favor vapor monitoring because it has the capability of providing an early warning of even small leaks in an UST system.

Vapor monitoring alarms are a function of several factors that include: 1) the type of vapor sensor, 2) alarm settings, 3) active leaks or spills, and 4) site conditions including soil moisture, temperature, backfill, and background vapor concentrations. Although a vapor alarm usually indicates a real condition of high hydrocarbons, it is not uncommon for the owner/operator to assume the system is not detecting a leaking UST, disable the system, and from then on to ignore it.

This study was conducted to examine how vapor readings changed over time, to understand why this occurred, and to evaluate implications for future leak detection efforts through vapor monitoring.

EQUIPMENT AND METHODS

[1]Senior Research Scientist, Veeder-Root Co., 125 Powder Forest Drive, Simsbury, CT 06070.

[2]Vice-President of Geosciences, On-Site Technologies, Inc., 1715 South Bascom Ave., Campbell, CA 95008.

There are four basic vapor monitoring approaches that include: 1) continuous aspirating, 2) continuous static/passive, 3) periodic aspirating, and 4) periodic static/passive. The first three are the more commonly used approaches for UST leak detection and examples of each of those are provided in this paper. The continuous aspirated method is represented by the vapor monitoring system manufactured by Arizona Instrument Corp. which uses a metal-oxide semiconductor (MOS) produced by Figaro U.S.A., Inc. to sense the vapor. The continuous static/passive method is represented by a Veeder-Root Co. system that employs an adsorption vapor sensor produced by Adsistor Technology, Inc. The periodic aspirating approach is represented by a portable device made by GasTech Inc. that employs a catalytic bead sensor.

All of the sensors collect data electronically and are based on providing total hydrocarbon vapor concentrations. However, as Portnoff [1] points out in this volume vapor sensors do not respond equally to the various volatile components of fuel and some may not respond at all to the highly volatile ones such as butane. Therefore, it is misleading to think that the concentrations in parts per million reported by a vapor monitor is equivalent to the more accurate results obtainable with a gas chromatograph.

The triggering of an alarm is also related to the type of fuel. For example, Raisanen [2] found that the vapor signal from gasoline was 44 times greater than from diesel fuel at the same leak rate. This was attributed to the lower volatility of diesel as well as the lower response of an MOS to diesel fuel's components. All of the following examples, except VM3, focus on vapor data around UST containing gasoline.

This paper reports on field data collected with a variety of sensors and four of the examples are from Santa Clara County in northern California, one is from Connecticut, and one from Texas. The regulations in Santa Clara County, as in all of California, require double-walled tanks for new or replacement tanks. As a result, their vapor monitoring experiences are limited to older installations.

VAPOR MONITORING DATA

A variety of vapor-data records have been selected in order to provide examples of different types of conditions. Fortunately, the majority of vapor records are monotonous graphs with few changes. Nevertheless, there is an uncomfortable number of graphs that have spikes, dips, and major fluctuations. These records are generally associated with specific incidents and the variations are explainable if enough is known about the site. The following examples of field data provide patterns of how vapor concentrations respond to various site conditions.

<u>VM1-Clean Site</u>

This site (Fig. 1) has only one 2,000-gallon (7,569 L) gasoline tank. It uses a continuous, aspirated vapor monitoring system with three probes and one that provides background readings. The vapor readings did not exceed 500 ppm over the entire reporting period. Although there is variation the monitoring record is typical for a clean, properly-operated facility.

<u>VM2 - Overfill Incident</u>

This site has one 550-gallon (2,081 L) gasoline storage tank and one vapor monitoring well. It was monitored on a monthly basis with a periodic-aspirating vapor monitor. An unknown amount of gasoline was spilled during filling of the tank between the March and April tests (Fig. 2). It is a strikingly simple example of what happens following an overfill.

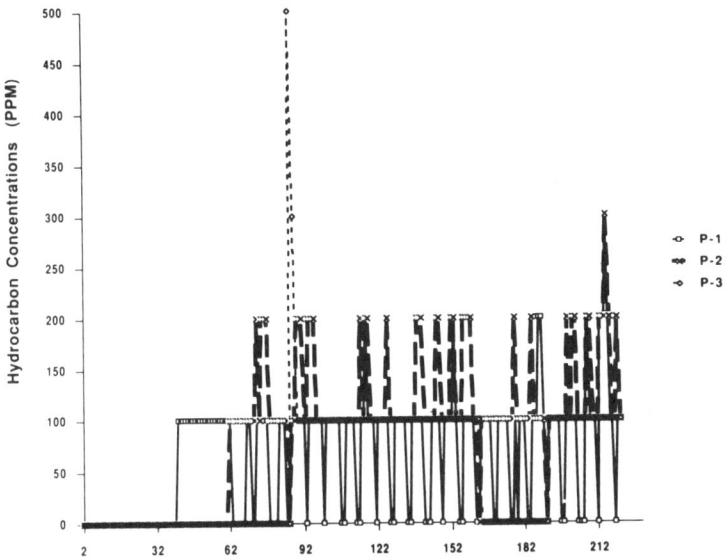

FIG. 1--Clean site for reporting period (186 days)

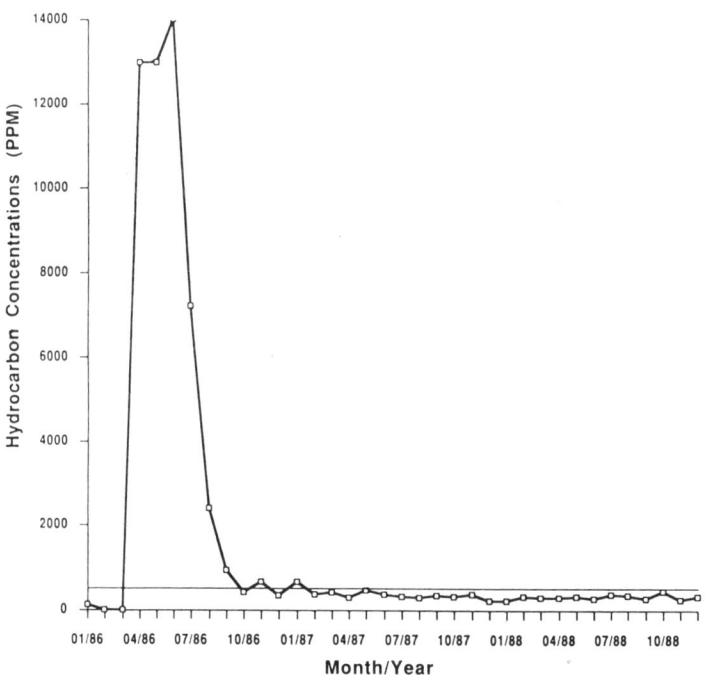

FIG. 2--Overfill incident

VM3 - Background Variation

This site (Fig. 3) has three 10,000-gallon (30,843 L) gasoline storage tanks. There are four observation wells that have continuous static/passive monitors. The water table is about 1.25 m below the top of the casing. Sensor #7 is .75 m deep in the casing while sensor #8 is only .5 m from the casing's top.

Since sensor #7 and #8 are both in the same well they are able to demonstrate the common occurrence of higher vapor concentrations at the bottom of a well caused by settling of the denser hydrocarbon vapors. Other influences may be the proximity of fuel in the capillary fringe to sensor #8 and/or the nearness of sensor #7 to the well cap and the atmosphere. Figure 3 also indicates the diurnal variation observed at some sites as vapor concentrations respond to temperature differences. Seasonal variations are thought to have a major impact on vapor monitoring with very low concentrations in the winter.

The water table at this site is somewhat high for vapor sensors and they have now been replaced by sensors that can detect floating product.

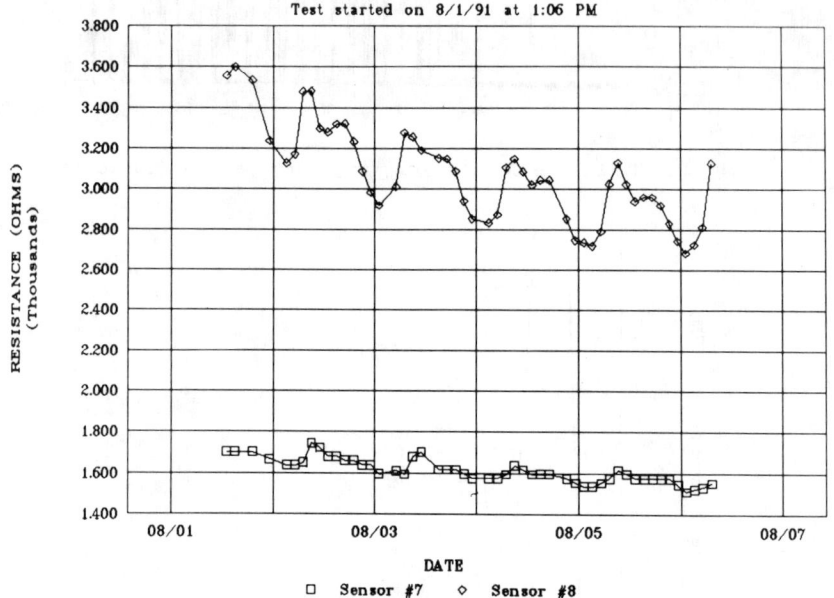

FIG. 3--Background variation

VM4 - Diesel Site

This site has one 1,000-gallon (3,784 L) diesel tank and was monitored by a continuous aspirated system. Detected vapor did not exceed 500 ppm over the entire reporting period (Fig 4). However, a rule of thumb is to give 100 ppm of diesel vapors the same attention as 500ppm of gasoline vapor. The peak of 500 ppm probably reflected a spill at the dispenser.

VM5 - Leaking UST

Figure 5 indicates a relatively short data record from a

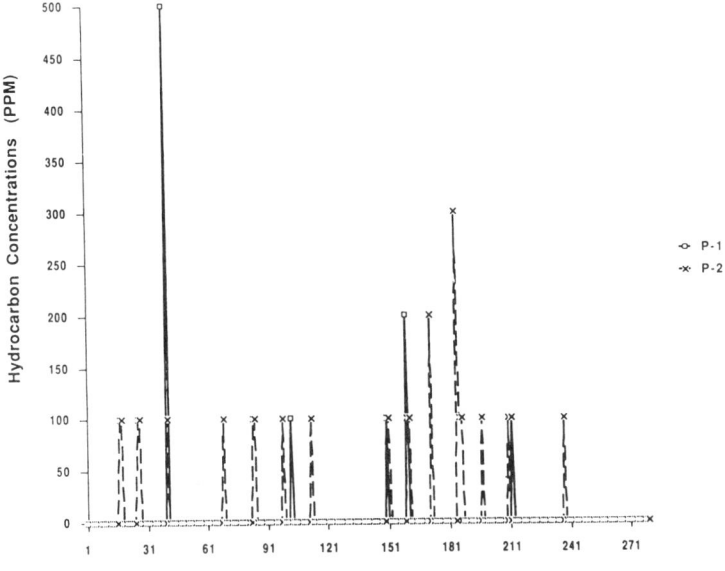

FIG. 4--Diesel site data for 280 days

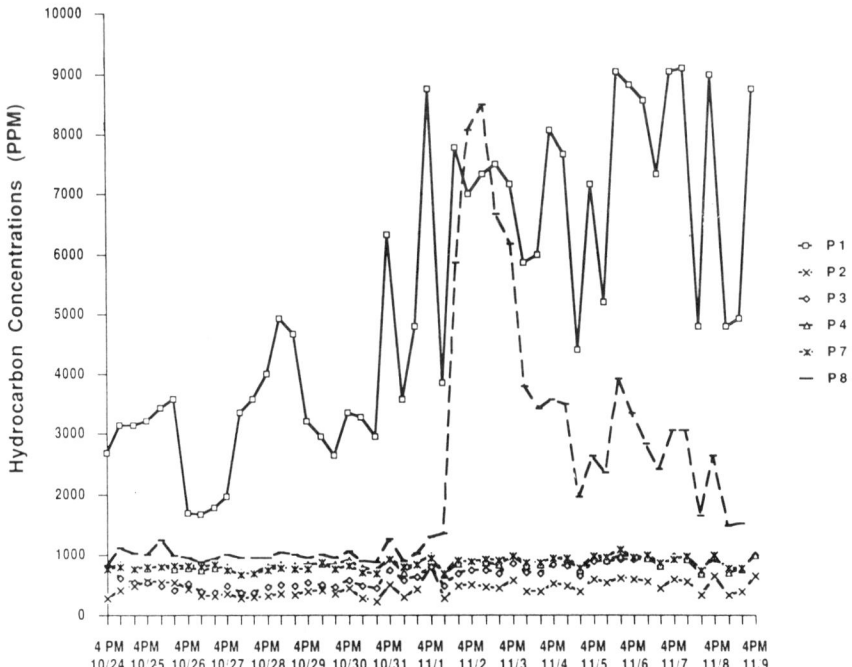

FIG. 5--Leaking UST

commercial filling station that has a continuous, aspirated, vapor-monitoring system. The results typify the detection of leakage in its early stages. The leakage is in the vicinity of Probe #1 and is probably minor. Evidence for that is that no other probes were affected. The high vapor event associated with Probe #4 is most likely the result of an overfill since the concentrations started falling back within eight hours after it peaked.

VM6 - Weeping Joint

This site has three gasoline tanks and four observation wells. The wells have adsorption vapor sensors and the readings as indicated in Figure 6 are way above the standard alarm of 10,000 ohms. The tanks and plumbing have passed the standard precision tightness tests. However, a slow weeping of gasoline from pipe joints by the pumps has been observed. This weeping is well below 0.05 gph but is enough to saturate the vapor sensors and make them useless until the joint is made tight.

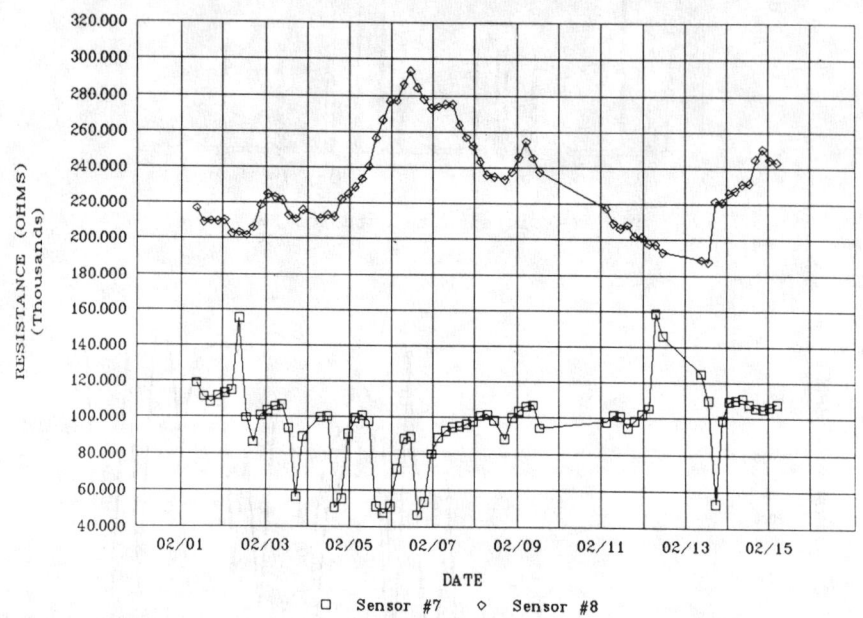

FIG. 6--Weeping joint

DISCUSSION

Site Assessment and Installation

Several considerations have arisen from the experiences of conducting vapor monitoring at UST sites. Initial installation of a vapor monitoring system is clearly important. Installation of a vapor monitoring system should be determined on a site-specific basis. For example, while it is possible to have a properly-operating, vapor-monitoring system with a shallow water table, it is usually preferable to employ groundwater monitoring if the water table is within the backfill.

There are some critical data that need to be collected to

determine the feasibility of vapor monitoring. The first question is
the depth to ground water at the site and if monitoring vapors in the
unsaturated zone is a feasible option. An equally important question
for the site is whether the subsurface permeability is adequate based on
the backfill (i.e. sand or gravel). A confounding factor at the site
can be if diesel fuel is to be monitored, particularly if tanks
containing gasoline and diesel are near each other. Another critical
piece of data is the pre-existing, hydrocarbon-vapor concentrations at
the start of monitoring. If the vapor concentrations are too high (i.e.
3000 ppm or higher) than the contamination should be assessed to
determine if the site can be adapted for vapor monitoring.

The contamination assessment can be subdivided into the following
steps: 1) sample and examine the soil around the fill risers of the
tanks, 2) inspect the product dispensers and the underlying soils, 3)
check the submersible pumps, sumps, and surrounding soils, and 4)
conduct a precision test of the tank and line systems. The
contamination assessment should conclude if there are leaks or
contamination to the extent that vapor monitoring is impractical. At
this point the site could undergo reconditioning to make it a suitable
candidate. The procedure might be to repair any leaking joints in the
piping, install overfill/spill protection devices, or even biotreat the
backfill by injecting air into the monitoring wells.

Data Analysis

Setting the alarm level is the primary and only data analysis tool
that most vapor monitoring systems employ. Once a system goes into
alarm that time may be stored in the system but no more than that. The
recommended alarm level is an arbitrary number set by the manufacturer.
For sites with high background concentrations installers use a guideline
such as four times the background value.

One of the impressions of this study was that although there are
definite improvements that could be made in vapor sensors many existing
systems can adequately detect hydrocarbons and provide relevant data.
However, most vapor sensing systems are only able to tell you what is
happening in real time. It is helpful if monitoring systems have some
memory so that records provide recent vapor history at the site. Data
interpretation is a critical necessity and this has been the greatest
weakness of most vapor monitoring systems. Nevertheless, this data
interpretation can be included in software that makes reliable leak
detection decisions automatically.

An important and encouraging conclusion is that the variability of
the background vapor concentrations appears to be minimal.
Nevertheless, information overload is a strong impression that one gets
from dealing with vapor monitoring data. This is particularly true at
sites with dirty housekeeping practices. The bad news is that although
the increase in vapor concentrations can be very rapid (i.e. hours or
even minutes), it may take a week or two before vapor readings begin to
decline and a spill can be identified with assurance. However, there is
some hope that the patterns of spills and leaks can be differentiated.
A leak tends to have vapor concentrations that trend upward over time
while spills are more likely to have a sharp spikes.

Substantial variability of hydrocarbon concentrations was found
between probes at a site. Although theoretically one vapor probe may be
sufficient to detect a leaking tank multiple probes are extremely
helpful. The additional probes can help identify the location of the
leak and explain its occurrence.

CONCLUSIONS

While these examples are only meant to illustrate representative
conditions, the data suggest several conclusions. An important finding
is that high vapor readings are indicative of specific occurrences

rather than being background noise. Therefore, if vapor monitoring is employed spills need to be minimized and overfill protection, in particular, should be installed. The time, effort, and aggravation caused by spills/overfills means that they must be avoided. Other common sources of subsurface contamination by fuel occurs when filters are being changed or maintenance is conducted on the dispensers or in the area of the tank pump.

The future of vapor monitoring around UST is difficult to predict. I think it is unlikely that vapor monitoring will ever become a strongly-favored option. Nevertheless, it is quite possible that as 1) less subsurface contamination occurs at sites, 2) consoles are equipped with memory as well as data analysis, and 3) new sensors appear on the market, vapor monitoring will gain favor and be seen as the most appropriate approach for specific applications. New installations are being constructed with catchment pans under the dispensers, overfill protection, and sump boxes around the pump. These features will reduce the amount of fuel that drains haphazardly into the soil causing high vapor concentrations. As a result, the data analysis function will be simplified.

Vapor monitoring is expected to be used in conjunction with internal monitoring methods to increase leak detection reliability. Vapor monitoring is particularly helpful at clean sites with single-walled tanks. It appears that some regulatory agencies will even require external monitoring (i.e. vapor or ground water) in conjunction with double-walled tanks.

As vapor monitoring improves with time it is expected to slowly gain more adherents. The primary reason being that as equipment and practices substantially reduce subsurface contamination at most sites, regulators may hold owner/operators to higher standards. Vapor monitoring is an ideal tool with which to ensure that those higher standards are met.

REFERENCES

[1] Portnoff, M., "Measurement and analysis of vapor sensors used for UST leak detection and site remediation", this volume.

[2] Raisanen, W., "Detecting low-level leaks of volatile petroleum products," <u>Sensors</u>, December 1989, pp 18-20.

Michael V. Martin,[1]

NEW VAPOR METHOD DETECTS AND LOCATES LEAKS FROM PIPELINES

REFERENCE: Martin, M. V., "New Vapor Method Detects and Locates Leaks from Pipelines," Leak Detection for Underground Storage Tanks, ASTM STP 1161, Philip B. Durgin and Thomas M. Young, Eds., American Society for Testing and Materials, Philadelphia, 1993.

ABSTRACT: This paper covers a new process that detects and locates leaks from pipelines using a patented sensor tube. The sensor tube is laid parallel to a gas or liquid pipeline to monitor for vapors from a release. The sensor tube membrane has a high permeability for vapor absorption and diffusion. A vacuum is pulled on the sensor tube with cycle time dependent on the vapor pressure of the leaking substance, distance the sensor tube is placed from the leak source, and the type, compaction, and moisture content of the soil. Vapors that have diffused into the sensor tube are transported to a pump/detection unit that monitors for an increase of vapor concentration along the length of the sensor tube. The system monitors for increases in vapors concentration over the background vapors present in the soil and around the pipe. The vapors are transported at a flow controlled rate allowing leak location to be calculated. Examples of actual applications and installation techniques are provided. The paper concludes with a summary of a third party protocol developed to evaluate this techniques capability.

KEYWORDS: leak detection and location, sensor tube, diffusion, absorption, vapor monitoring

Vapor Monitoring Wells are an accepted method to monitor for leaks from underground storage tanks and pipe [1]. This paper discusses a method that is similar to well leak detection technique, yet superior for pipelines for reasons to be defined.

Vapors from a leak source migrate through the soil. The time required for a leak to migrate a known distance is dependent on the vapor pressure of the leaking substance, the leak rate, and finally, soil type, compaction, and moisture content. Therefore, it is logical that the closer the vapor collection device is to the leak source the shorter the time required for vapors to be collected. This leak detection and location method is based on the premise that the sensor tube acts as a continuous row of wells that can all be automatically sampled with one central pump/detector. The system precisely locates the leak. The benefit of this system is earlier detection and location of smaller leaks to minimize product loss to the environment.

Dr. Wolfgang Issel developed Leak Alarm System for Pollutants, LASP, with the support of the German Ministry of Research and Technology [2] to protect groundwater and other environmentally sensitive zones.

---

[1] Director, Environmental Monitoring Division, Teledyne Geotech, 3401 Shiloh Road, Garland, TX 75041.

LASP SYSTEM OPERATION

Typical LASP operation is illustrated below in (FIG. 1). The LASP system consists of the patented sensor tube permanently installed near the leak source. The sensor tube contains atmospheric air that is filtered and dried. The sensor tube is installed along the area to be monitored in possible continuous lengths of over 16 kilometers. A leak medium that comes in contact with the wall of the sensor tube, forms a vapor plug inside the tube by diffusing through the wall of the tube. The leak medium can be in the form of a gas, liquid, vapor, or constituent dissolved in water. Detectable substances include a wide range of gases, hydrocarbon liquids (jet fuels, diesel fuels, gasolines) and vapors, halogenated hydrocarbons, land fill gases, water vapor and many others. To detect and locate the leakage, the air column containing the vapor plug is transported at certain intervals, by a pump/detector unit at the end of the sensor tube. The signal from the gas sensor stands near zero unless a vapor plug from a leak is passed by. The signal produced is called a leakage peak and is proportional to the concentration of the vapor from the leak medium at the sensor tube surface. The system is self tested with every cycle by automatically injecting a small gas volume into the end of the sensor tube. This gas volume creates an end peak that precisely marks the end of the sensor tube for accurate leak location.

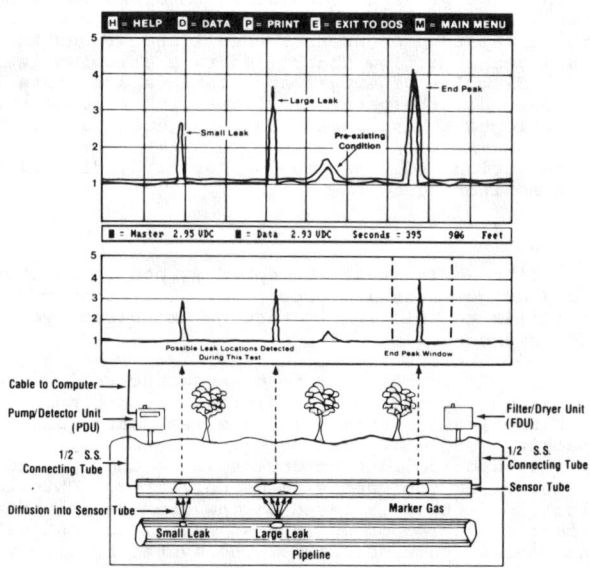

FIG. 1--LASP System Operation

THE SENSOR TUBE (COLLECTION DEVICE)

The patented sensor tube (FIG. 2) parallels the pipe to be monitored, preferably within 1 meter of the pipe circumference, and along the length. The sensor tube diameter is 1.27 centimeters. The sensor tube is impervious to water and is made up of a rigid PVC inner core with holes perforated in it. A thin Ethylene Vinyl Acetate (EVA) membrane is extruded over the PVC inner core. EVA is a plastic membrane with a high permeability for vapor molecules. The life expectancy of the sensor tube is equal to the life of the pipeline. A polyethylene braid is extruded over the membrane to protect if from damage. The tube design effects the diffusion of the leak medium and the transportation of the vapor plug inside the sensor tube. Variables that effect the diffusion process into the sensor tube are the number of perforated holes and the membrane material and wall thickness.

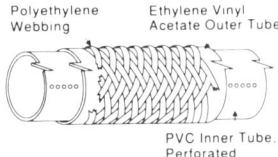

FIG. 2--LASP Sensor Tube

The absorption, diffusion and transportation process are shown in (Fig. 3). In A, a vapor cloud from a leakage migrates towards the sensor tube. The concentration of this vapor cloud is dependent on the leak rate and vapor pressure/temperature of the leaking substance. The soil parameters also influence the diffusion of this vapor cloud. A sandy soil or gravel has much quicker diffusion than a clay or tightly compacted moist soil [3]. In B the vapor comes in contact with the sensor tube and will be absorbed by the outer layer. The molecules migrate through the outer layer by diffusion and in C vapors move into the air volume inside the sensor tube through small holes in the inner layer. The wall diffusion is also dependent on the leak medium vapor pressure/temperature. Diffusion rate increases with the vapor pressure/temperature of the leak medium. There is a difference in the diffusion rate into the sensor tube for a permanent light gas and a heavy vapor. A light gas diffuses through the outer layer with minimal molecular structure change. A heavy long chained or aromatic vapor, causes the membrane to swell slightly by strong interaction with the outer layer material. In D the vapor plug is now transported in the sensor tube. In E the vapor plug size is increased during transportation and the dual wall sensor tube minimizes reabsorption of the molecules. F illustrates single layer tube vapor plug reabsorption. After only 100 meters the vapor slug is completely reabsorbed.

The velocity at which the vapor slug is transported is important. With a normal velocity of 1 meter/second the reynolds number is in the transition zone which is optimum. If the velocity is much higher you need a high pressure differential which creates mechanical stress. Lower velocities in the laminar flow range result in dilution and elongation of the vapor slug. Dilution reduces the vapor concentration and elongation decreases the location accuracy. With lower velocities, running time is longer, and there is more time for reabsorption which decreases the leakage peak.

PUMP/DETECTOR UNIT

The pump/detector unit houses the vacuum pump, vapor sensors, flow controller, relays, power supplies, and communication devices to interface to the computer. Sensors can be selected depending on the application. For applications where high methane background exists, sensors with minimal methane sensitivity are available. A mini infrared (IR) sensor [4] can be used that only responds when a specific vapor is present which eliminates background interference from other vapors. Other sensors that are effective are metal oxide non methane sensitive

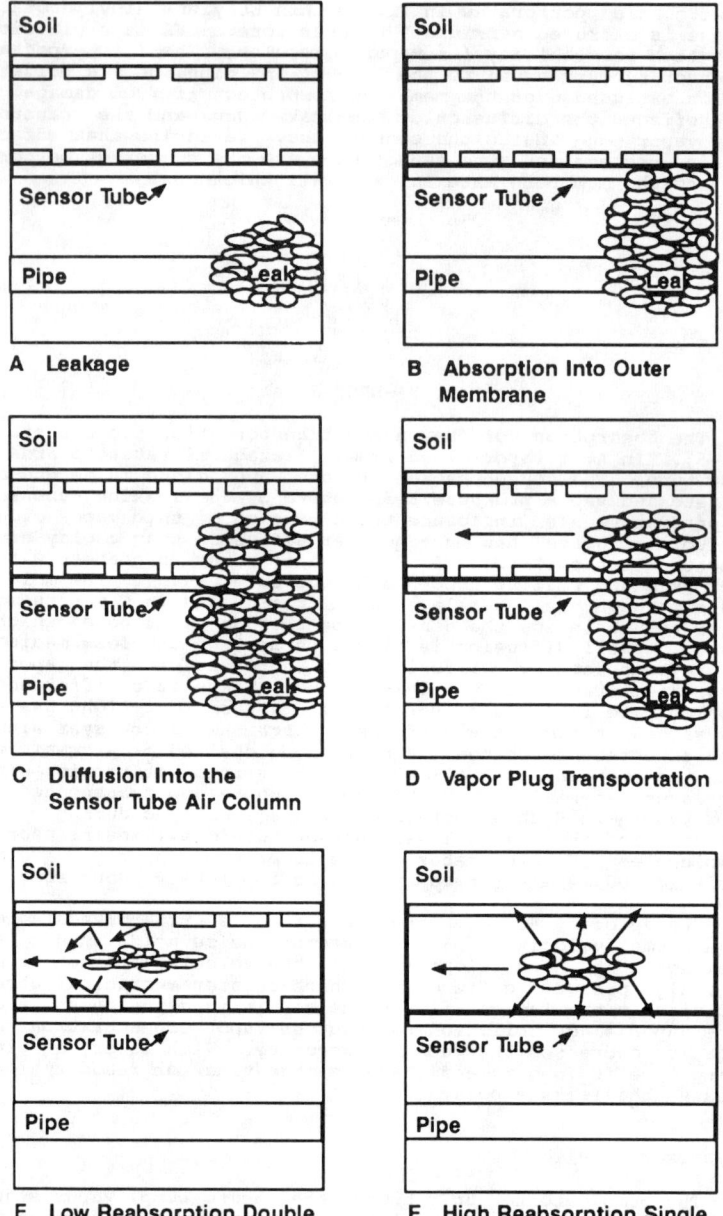

FIG. 3--Absorption, diffusion, and transportation of vapor slug.

sensors [5][6][7]. LASP can be used for any vapor that can diffuse into the sensor tube and be detected by an on line real time analyzer.

The pump/detector unit vacuum pump pulls the air column in the sensor tube past the sensor. The voltage output of the pump/detector Unit Sensor is compared to the previous background level. If this output exceeds the configured percent above the background at any point along the sensor tube, the system alarms, indicating a possible leak has occurred. It is important to note a percent set point allows the system to accommodate varying background over the length of the sensor tube.

LASP automatically locates the leak by calculating the time it takes the leak peak to reach the pump/detector unit and by controlling the flow rate (velocity) of the air column. The velocity is multiplied by the time which equals the distance. The end peak is used as a reference measure to decrease the location error. In practice it is possible to locate leaks within +/- 15.2 meters on a 6.2 kilometer pipeline.

FILTER/DRYER UNIT END PEAK GENERATOR

The filter/dryer unit is installed above ground at the end of the sensor tube to prevent water vapor from condensing inside the sensor tube and to eliminate the possibility of pulling polluted vapors into the tube from the atmosphere that would cause false alarms. The filter/dryer unit consists of a carbon filter, silica gel dryer, and means to generate an end peak. During the configured diffusion time, the filter/dryer unit generates a gas end peak which marks the precise location of the end of the tube and monitors tube integrity each time the system performs a leak test.

PC CONTROL UNIT

A pc controls the system and stores each leak test file by date and time. The pc is also used to display the stored files. The crt display is a split screen. The lower screen shows the entire section up to 3.1 kilometers. Possible leak locations are marked by an arrow. The upper display shows up to 365 meters of the lower display with higher resolution. The pc controls continuously monitors the alarm functions. Dry contact closures are available for remote alarms.

PRACTICAL APPLICATIONS

Practical Applications include underground gas (including dangerous H_2S [8][9]) and liquid pipelines, hydrant fuel systems, storage tanks, and underwater applications below silt level. Other applications include remediation progress monitoring, landfills, and sewer pollution monitoring.

This technique can also be used for above ground liquid pipeline applications with the sensor tube installed under the pipeline at the apex of the circumference. Leaking fluid will contact the sensor tube and the absorption and diffusion process is immediate with liquid contact.

INSTALLATION TECHNIQUE

<u>New Pipeline Installations Technique</u>

<u>On shore</u>--The sensor tube should be placed as close to the leak

source as possible. The simplest location is on top of the pipeline at the apex of the circumference. Because the system detects vapors from a leak, and is not dependent on liquid contact, placement of the sensor tube is not critical. LASP can detect leakages from more than one pipeline by placing the sensor tube between the pipes. The maximum center line distance between pipes should not exceed 3 meters.

Offshore--The sensor tube can be attached to the pipe as it is lowered to the sea bed.

Existing Pipeline Installation Technique

On Shore--LASP has been successfully installed where soil conditions permit utilizing a vibratory cable layer. An illustration in (FIG. 4). This technique allows for a fast one step efficient installation. The tube is inserted into a slotted plow blade that is pulled through the ground by a tractor as it vibrates. The vibration reduces friction and increases the speed that the plow blade moves through the ground. The pipeline is located with a metal detector and marked. The tractor follows these marks and installs the sensor tube to one side of the pipeline. This eliminates the chance of damage to the pipeline from depth variation caused by soil erosion. No pipeline damage has occurred using this technique. Installation rates are average 200 meters/hour.

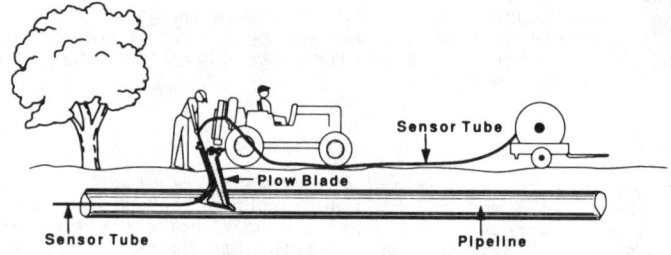

FIG 4--Illustration of the vibratory cable layer installation technique.

Off Shore--It is possible to install LASP on an existing offshore pipeline after location at depths reachable by divers. A jet sled is used to move silt away while divers attach the sensor tube to the pipeline.

LEAK DETECTION AND LOCATION EXAMPLE

This application was a new pipeline construction in 1978 crossing the Rhine River near the city of Worms, Germany. Total length of the sensor tube was 3,000 meters. The LASP system is still operating on this pipeline transporting ethylene gas. A diagram of the application is shown in FIG. 5. Several months after starting operation of the pipeline pressure was increased to normal operating pressure. Two leakages were detected and located. A group of gate valves was identified as the location of the leakages. These gate valves were dug up and leakages verified. The ethylene gas leakage rate from each valve was about 11 liters/hour. The pipeline also had volume balance and pressure drop systems monitoring for leaks that did not detect these small leaks.

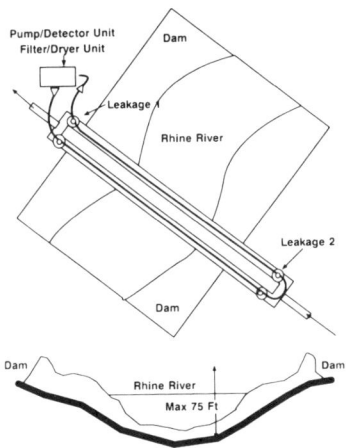

FIG. 5--Leak example diagram of the ethylene pipeline installation at Worms, Germany.

CONCLUSIONS AND SUMMARY OF LASP PERFORMANCE EVALUATION

At this time an EPA protocol is not adopted for third party evaluation of a LASP type system. EPA protocols do exist for the sensors used in vapor systems [10]. Evaluation of collection devices such as LASP sensor tube are critical to the overall evaluation of a systems ability to detect and locate a low level leak and should be part of the protocol. In March 1991 Jones and Neuse, Environmental Engineering and Consulting was retained to develop and conduct a worst case protocol for evaluation of LASP [11].

JP-5, a very low vapor pressure jet fuel was selected as the test fuel. The leak rate was 0.076 liters/hour. The soil was sandy lean clay. The leak was detected within 4 days with only 7.3 liters of jet fuel in the soil. Jones and Neuse summarized the performance evaluation conclusions as follows:

"Results of the testing performed by JN to evaluate the performance of the LASP system confirm that the system is capable of detecting organic vapors that could originate from leaking UST and piping systems in a setting that simulates a UST and pipeline back fill. The system was evaluated under approximated "worst case" circumstances by using a relatively low permeability back fill material and a relatively low volatility hydrocarbon product. The LASP evaluation testing confirmed the relationships that would be anticipated from a properly functioning system. The following generalizations can be derived from the results of the LASP evaluation testing:

The highest concentrations of organic vapors were measured in LASP monitoring installations that were closest to the fuel leak and consistently declined with increasing distance form the source of the fuel leak.

Organic vapor concentrations measure in LASP monitoring locations generally increased over time as additional volumes of fuel were continuously released.

Organic Vapor concentrations measured by the LASP system were delayed in time and reduced in magnitude by back fill conditions with a higher moisture content, higher density, decreased porosity, and lower permeability.

The arrival time of the hydrocarbon peaks were consistent during the evaluations and reflected the monitoring system configuration.

Organic vapor leak detection systems are dependent upon an accurate and reliable organic vapor sensing device."

REFERENCES

[1] Office of Underground Storage Tanks, United States Environmental Protection Agency, EPA/530/UST-89/012, Detecting Leaks Successful Methods Step-by-Step, EPA, Washington, D.C., 1989.

[2] Issel, Dr. Wolfgang, "Leakage Detection System for Pipelines (LASP)", Report to the Rhein-Westfalia Technical Supervisory Committee, Germany, October 8, 1981.

[3] Office of Underground Storage Tanks, United States Environmental Protection Agency, EPA/530/UST-89/012, Detecting Leaks Successful Methods Step-by-Step-Chapter VI-Vapor Monitoring, EPA, Washington, D.C., 1989.

[4] Downie, Ron, "Mini IR Sensor", Teledyne Analytical, City of Industry, California, October 1991.

[5] Hinchee, R.E., Nielsen. B.J., Wickramanayake, G.B., "Evaluation of Liquid and Vapor Monitoring Devices for Underground Storage Tank Leak Detection", Defense Technical Information Center, Cameron Station, VA, 1991.

[6] Portnoff, M.A, Grace, R. Guzman, A.M., Hibner, J., "Measurement and Analysis of Adsistor and Figaro Gas Sensors Used for Underground Storage Tank Leak Detection", Carnegie Mellon Research Institute, Carnegie Mellon University, Pittsburgh, PA, 1991.

[6] Portnoff, M.A, Grace, R. Guzman, A.M., Hibner, J., "Measurement and Analysis of Vapor Sensors Used for UST Leak Detection and Site Remediation", Carnegie Mellon Research Institute, Carnegie Mellon University, Pittsburgh, PA, 1991.

[8] Sperl, J.L., "System Pinpoints Leaks on Point Arguello offshore line", Oil and Gas Journal, Volume 89, No.36, page 47-52, September 9, 1991.

[9] Mannan, Mahboobul, Pfenning, Dwight B., Zinn, Dale C., "SOUR-GAS PIPELINE-1: Risk-analysis procedures ensure system safety", Oil and Gas Journal, June 3, 1991, Vol,89, No.22, page 83-87, "SOUR-GAS PIPELINE-Conclusion: Line, Weather conditions among variables to determine public risk", Oil and Gas Journal, Volume 89, No. 23, June 10, 1991.

[10] United States Environmental Protection Agency, Standard Test Procedures for Evaluating Leak Detection Methods: Vapor Phase Out of Tank Product Detectors, EPA/530/UST/90/008, Washington DC, March 1990.

[11] Onjanow, A., Talley, A., LASP Performance Evaluation Report, Jones and Neuse, Inc., Environmental and Engineering Services, Austin, TX, 1991.

Glenn M. Thompson[1], Randy D. Golding[2]

Pipeline Leak Detection Using Volatile Tracers

REFERENCE: Thompson, G. M. and Golding, R. D., "Pipeline Leak Detection Using Volatile Tracers," Leak Detection for Underground Storage Tanks, ASTM STP 1161, Philip B. Durgin and Thomas M. Young, Eds., American Society for Testing and Materials, Philadelphia, 1993.

ABSTRACT: A method of leak detection for underground storage tanks and pipelines adds volatile tracers to the product in the tanks and analyzes the surrounding shallow soil gases for tracer vapors. This method has several advantages: the success of the test is not limited by the size and structural design of the vessels, tanks can be tested at any fill level without taking the tank out of service, the location of a leak along a pipeline is clearly marked by the location of the tracer, and liquid leaks as small as 0.2 liters per hour (lph) can be detected. A limitation is: the backfill material must have some degree of air permeability in the zone above the water table. Several field tests document the success achieved using this method. A tracer leak detection system was installed at Homestead AFB after several other testing methods failed to locate a leak at a valve pit location along approximately 4 kilometers of fuel transfer piping. The leak was detected to the side of the valve pit at a depth of approximately 2.5 meters below the ground surface. Another installation at Edwards AFB involved the collection of 415 soil gas samples along approximately 3,050 meters of 15.25-centimeter fiberglass pipeline. Fourteen separate leaks were detected.

KEYWORDS: tracer, soil gas, underground storage tanks, pipelines, air permeability

INTRODUCTION

A method utilizing inert chemical tracers has been developed for leak detection of underground storage tanks and pipelines. The tracer, a volatile chemical concentrate, is mixed with the product inside the tank or pipeline. The starting concentration of tracer is 1 to 10 parts per million (ppm) in the fuel. If the product leaks into the soil outside of the fuel system, the tracer escapes from the liquid by evaporation. The tracer vapors are released into the soil and migrate in all directions throughout the soil porosity. Hollow probes are placed in the soil near the tank or pipelines to collect the tracer vapors that appear in the event of a leak. The vapors are collected from the soil and analyzed for the presence of tracer by means of an extremely sensitive chromatographic measurement. The tracer is detectable at the low-parts-per-trillion level making it possible to detect very small leaks with only 10 ppm or less of tracer in the

---

[1] Ph.D., President, Tracer Research Corporation, 3855 North Business Center Drive, Tucson, AZ 85705

[2] Ph.D., Director Special Projects, 3855 North Business Center Drive, Tucson, AZ 85705

product. This low tracer concentration is not known to have any
discernable impact on the properties of any product.

RANGE OF APPLICATION

Leak detection with volatile tracers has been successfully applied to
underground and above ground tanks, pipelines and a variety of
plastic-lined impoundments including landfill liners since the
mid-1980s. Unlike volumetric, vacuum and pressure tests, the size and
structural design of the vessels being leak tested are typically not
factors limiting the applicability of the tracer leak detection
methods. For example, the sensitivity and accuracy of volumetric tests
and pressure tests are severely impacted as tank or pipeline size
increase because thermal effects develop proportionally with vessel
size and mask progressively larger leaks. Vacuum testing can only be
applied to structures that will tolerate external pressure.

Tracer testing depends only on the detection of a chemical outside of
the tank after it has been placed on the inside. The tracer will be
carried through apertures (leaks) in the vessel wall either with the
escaping fluid under the normal operating conditions (advective flow)
or, if the vessel is empty, the tracer vapors will escape through the
hole by molecular diffusion. Tracer Research Corporation has verified
that both of these processes, advective flow and molecular diffusion,
are effective for transporting tracer through holes in underground
structures. These tests were performed as part of the third party
verification of the ***Tracer Tight®*** Leak Detection System [1,2]. In
these tests, detection of liquid leaks of 0.2 lph and passive leakage
by vapor diffusion through holes of 0.08 millimeters in diameter were
verified. This ability to detect either vapor or liquid leakage makes
the method particularly useful for tank testing because it allows the
tank to be tested at any fill level and without taking it out of
service.

An additional aspect of the tracer method that makes it uniquely
appropriate for pipeline testing is the fact that the location of the
tracer in the soil along the pipeline indicates the location of the
leak. The ability to locate the leak is commonly the principal reason
for performing the tracer leak test on pipelines. In many cases, the
tracer test is initiated after the pipeline has failed a pressure test.

LIMITATIONS

The principal limitation to the tracer leak detection method is the
permeability of the backfill around the pipe. In order for the leak to
be detected, the backfill must have some degree of air permeability in
the zone above the water table. This air permeable zone must be at
least 0.6 meters thick. The fact that the pipeline itself may be below
the water table presents no special problem if the pipeline is leaking
a petroleum product. Most petroleum fuels and solvents have a density
lower than water and will float to the surface of the water table if
released underwater. Once the escaping product reaches the water table
surface it contacts unsaturated soil. At this point the tracer
chemical is able to vaporize out of the product and diffuse into the
air-filled pores of this soil. At least 0.6 meters of unsaturated soil
is required for the tracer to diffuse laterally a sufficient distance
to be detected by sampling probes placed up to 3 meters away. A
thinner unsaturated zone would reduce lateral saturation and cause
excessive tracer loss to the above ground atmosphere. The minimum

permeability of the soil required for this amount of lateral diffusion is typically about 1 darcy or $1 \times 10^{-8} cm^2$. Most soil types, including clay rich soils, will have adequate permeability for tracer leak testing if the soil is moderately well drained. The vast majority of pipeline and tank locations are in moderately well-drained soils. Topographic low points, where the soil is perennially swampy and standing water is commonly on the surface, are the problem areas. In these areas of complete water saturation, fuel will commonly migrate to the ground surface and become visually detectable as sheen and puddles on the ground surface.

HOMESTEAD AFB CASE STUDY

Homestead AFB was the first military base to install a tracer leak detection system. It was installed along approximately two and a half miles of fuel transfer piping and around fifty-two 200,000-liter underground tanks. The system consisted of leak detection hose along the pipeline and probes around the tanks. The leak detection hose is a specially designed hose with a very low wall permeability. When the hose is placed under vacuum, vapors can be extracted evenly along its entire length. It is employed in maximum lengths of 175 meters. With one end plugged, air is drawn from the other end to sample vapors along the pipeline. If tracer is detected in the air drawn from the hose, the pipeline is presumed to be leaking somewhere within the 175-meter hose interval. The leak location is then determined by placing probes every 3 to 6 meters along the pipeline until the leak is located.

Homestead AFB was selected by Tactical Air Command as the first place to apply the leak detection system because that particular base was plagued by the periodic appearance and disappearance of fuel at a valve pit location along the piping. The base personnel were divided in their opinion as to the origin of the fuel. Some thought it represented an active pipeline leak and others thought the fuel was from spillage that may have occurred years earlier. Several attempts to find the leak in the two years preceding the installation of the leak detection system had failed. The most significant effort involved digging up about 30 meters of pipeline on the north side of the valve pit where the leak appeared. However no leaks were found. After installation of the leak detection system, tracer was detected on the opposite side of the valve pit from where the pipeline had been uncovered.

The distribution of tracer along the pipeline above the leak is illustrated in Figure 1. The tracer had been in the fuel system for approximately three weeks before the soil gas was tested. The tracer was detectable in the shallow soil gas for at least 6 meters along the pipeline. The highest tracer concentration is typically over the leak location, although occasionally it may be displaced as much as 1.5 to 3 meters laterally from the leak. In this instance, the leaking pipe was approximately 2.5 meters below the ground surface. The discovery and repair of the pipe was made especially difficult for the following reasons. There were four pipelines occupying the pipe ditch in this area. The two upper pipes were side by side about 1.5 meters deep and the two lower pipes were positioned directly below them at a depth of approximately 2.5 meters. The leak was in one of the lower pipelines. A high voltage electric line crossed the pipe ditch about 3 meters to one side of the leak location presenting a hazard to the digging operation. The depth to water at this time was approximately 0.8 meters. Consequently the digging was made difficult by a constant flow of water and fuel into the excavation. Special water removal and water treatment equipment had to be brought to the site to drain the excavation and separate the fuel from waste water. As a final point, the soil, which was a coral substrate, was very hard to dig through. Due to the hazard posed by the electric line, the fuel, and the pipes, the digging was done primarily by hand, which meant jack hammers and shovels -- a relatively slow process.

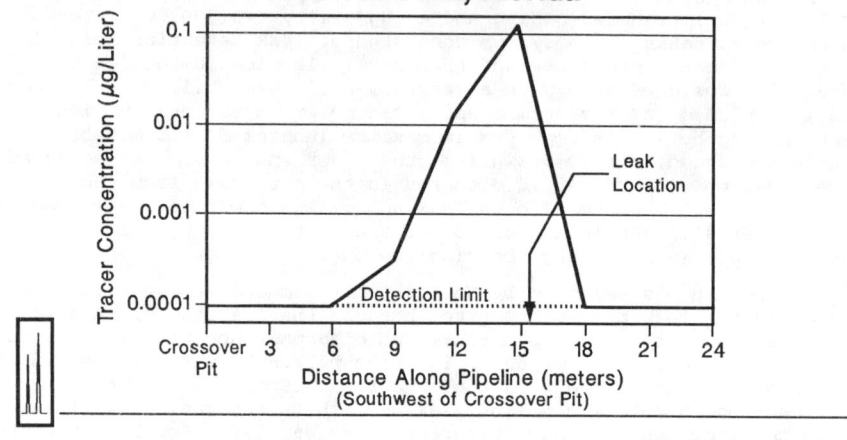

Figure 1
Tracer Concentration in Soil Gas Along JP-4 Pipeline
2.5 Meters Below Ground Surface, 1.7 Meters Below Water Table
Homestead AFB, Florida

The leak was eventually detected in direct correlation to where the highest concentration of tracer was detected. The excavation required approximately eight days. The reason it took so long was partly because the ditch was made unnecessarily long (due to the hope of finding the leak in the upper pipes) and partly because of the obstacles mentioned earlier.

The hole in the pipe appeared to have resulted from galvanic corrosion in the bottom side of one of the pipes. No information was obtained to explain its occurrence in that particular place. The 2.5-meter pipeline was repaired with a large band clamp and the excavation was refilled. The fuel system remained in normal use though the entire operation.

This information was taken from the investigative report, 3TFW/LGCV, provided by Tracer Research to Homestead Air Force Base, June 1988.

EDWARDS AFB CASE STUDY

At this site approximately 3,050 meters of a 15.25-centimeter fiberglass pipeline used for JP-4 was tested. Approximately 18,000 grams of the tracer was added to the bulk storage facility, bringing the initial concentration of tracer in the system to approximately 10 ppm. The tracer labeled fuel was distributed throughout the entire system and testing was begun 21 days later.

Four hundred and fifteen soil gas samples were taken and analyzed along the pipeline. During the course of the investigation, tracer was detected in 27 samples corresponding to 14 separate leaks. Figure 2 is a graphical presentation of the leaks. On the Y-axis is the concentration of tracer that was detected. The X-axis shows the distance plotted from the bulk storage area. The graph shows the location of the 14 leaks. The largest "peaks" on the graph indicate the locations with the greatest leakage. Leak number 11 was the largest leak while number 14 was the smallest.

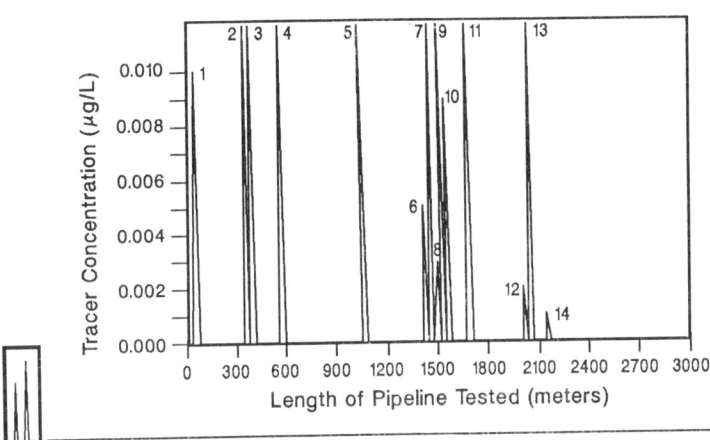

Figure 2
Tracer Concentration in Soil Gas Along Pipeline

Figure 3 is a graph of the hydrocarbon vapor concentrations versus the distance from the bulk storage area. This plot shows that the measured hydrocarbons do not correspond to the leak locations in all cases. The largest leaks detected show significant levels of the total hydrocarbons, but there are several areas where significant levels of hydrocarbons are detectable where no tracer was evident. The area approximately 2100 meters beyond bulk storage is a fuel handling area. In this area the fuel is loaded and down-loaded from the airplanes. It is very likely the hydrocarbons detected in this area are due to spillage.

This pattern of numerous small leaks along fiberglass piping is typical. The leaks are associated with poorly made joints, very minor damage due to rough handling of the pipe, and very small manufacturing flaws in the pipe. This mode of failure contrasts with that of steel pipe which tends to remain leak free until corrosion causes it to fail. The corrosion leaks in steel pipe, however, are typically larger than the individual leaks caused by the flaws in the fiberglass pipe.

Figure 3
Hydrocarbon Concentration in Soil Gas Along Pipeline

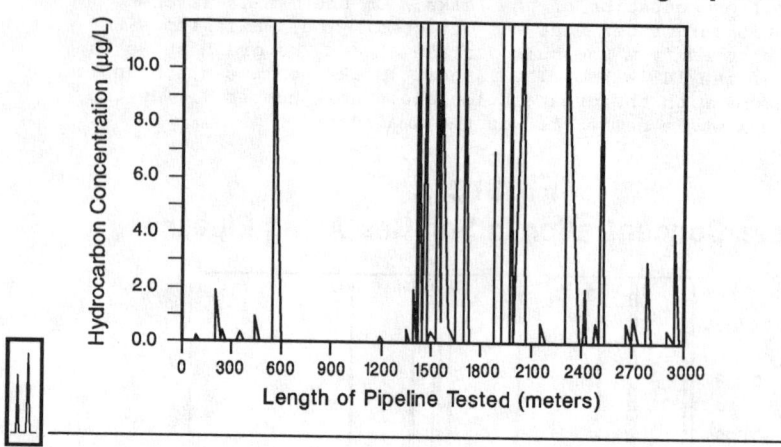

This information is extracted from the report of an investigation conducted for the Army Corps of Engineers by Tracer Research Corporation at Edwards AFB, CA, 1990.

CONCLUSIONS

The tracer leak detection method circumvents all of the problems associated with conventional leak testing of tanks and pipelines. The only requirement for the successful employment of a tracer leak detection system is that the soil have some degree of air permeability in the zone above the water table.

Third party evaluations and testing at major U.S. Air Force Bases have confirmed the following: Tracer leak detection can detect and locate very small leaks in large or small tanks or pipelines without interrupting the service of the fuel system.

REFERENCES

[1] Performance Evaluation of the Tracer Tight Leak Detection, Final Report. October, 1990. Prepared by Ken Wilcox Associates, 1312 521st St., Blue Springs MO., for Tracer Research Corporation

[2] Performance Evaluation of Tracer Tight Detection of Passive Ullage Leaks by Control Strategies Engineering, Dec. 1991, for Tracer Research Corporation

Regulations and Standards

Thomas M. Young[1]

HOW WELL DO LEAK DETECTION METHODS WORK?: PRELIMINARY RESULTS FROM THE EPA TEST PROCEDURES

REFERENCE: Young, T. M., "How Well Do Leak Detection Methods Work?: Preliminary Results from the EPA Test Procedures," Leak Detection for Underground Storage Tanks, ASTM STP 1161, Philip B. Durgin and Thomas M. Young, Eds., American Society for Testing and Materials, Philadelphia, 1993.

ABSTRACT: In 1990, EPA released a series of "Standard Test Procedures for Evaluating Leak Detection Methods" to define consistent, rigorous procedures for determining the performance of underground storage tank leak detection methods. Each test procedure outlines one or more ways to evaluate the performance of a class of leak detection devices. Liquid- and vapor-phase detectors are tested for their ability to detect the presence of petroleum components in controlled laboratory settings. Pipeline leak detectors, tightness testing methods and automatic tank gauging systems are tested for their ability to reliably differentiate between non-leaking systems and systems with simulated leaks of specified sizes. Statistical inventory reconciliation methods are tested using real inventory data from tight tanks, including some simulated leaks. The results of these preliminary evaluations are discussed and compared. The test procedures allow manufacturers to demonstrate the effectiveness of their methods and help tank owners to ensure that the equipment or service they purchase complies with applicable regulatory requirements.

KEYWORDS: leak detection, performance evaluation, standard test method, tank testing, groundwater monitors, vapor monitors, automatic tank gauging, pipeline leak detectors,

In preparing to issue regulations governing underground storage tanks (USTs) EPA discovered that, although hundreds of brands of leak detection were already in commercial use, most performance claims were based on little or no evidence. EPA studies of the important general classes of leak detection technology (e.g., groundwater monitoring, vapor monitoring, tank testing) showed that each was effective in many cases, but that no one technology was best for all types of UST sites. In the final regulations (40 CFR 280 Subpart D) the agency allowed each method provided that it met a performance standard specific to that method. For example, groundwater monitoring devices must be able to detect 1/8 inch (0.32 cm) of floating product within a well. As an additional encouragement to innovation, the regulation included a provision allowing the use of any method provided that it can detect a leak of 0.2 gallons per hour (0.76 L/h) or a release of 150 gallons (567.8 L) in a month with a probability of detection of 0.95 and a probability of false alarm of 0.05.

---

[1]Department of Civil and Environmental Engineering, University of Michigan, Ann Arbor, Michigan 48109-2125.

The performance standards are intended to allow only the most effective leak detection methods to be used to meet the regulatory requirements. Since a method's estimated performance is dependent on the rigor of the conditions under which it is evaluated, a performance standard will fail to achieve its goal unless minimum standards for an evaluation are established. Unfortunately, no national consensus code exists for evaluating leak detection equipment and manufacturers have followed a wide range of procedures to test their methods. To address this need, EPA developed standard test procedures that equipment manufacturers (or consultants working for them) could use to evaluate their equipment [1-7]. Separate test procedures have been developed for each of the following methods: liquid-phase and vapor-phase product detectors, volumetric and non-volumetric tank testing methods, automatic tank gauging systems, statistical inventory reconciliation methods, and pipeline leak detection systems.

Each document describes a series of tests to determine a method's ability to distinguish a leak (or the presence of product) from a tight tank (or the absence of product). Data collected during the evaluation are recorded on forms provided in each document. Evaluation results are summarized on a short, standardized reporting form. Leak detection manufacturers are responsible for distributing the results sheet to purchasers and state or local regulators who request the results for method approval. The forms allow tank owners or regulators to determine if a particular brand of leak detector meets the federal regulatory standards. The forms also include other important information, such as limitations of the evaluation results or key procedures for using the method.

EPA's standard test procedures for leak detection methods are only minimum standards for an acceptable evaluation. Alternative evaluation procedures may be developed by nationally recognized associations such as ASTM or by independent third-party laboratories who certify that the procedure is at least as rigorous as EPA's test procedure. To be as rigorous as EPA's procedures, an alternate test procedure must include physical testing of a full-sized version of the leak detection equipment. At least as many test replications must be conducted under as great a range of experimental conditions.

The test procedures differ significantly for leak detection systems that measure properties inside the tank system to detect a leak and those that detect the presence of product outside the tank once it has escaped. The remainder of this paper describes the test procedures for external methods, internal methods and pipeline leak detectors in greater detail and presents some preliminary results from the application of these test procedures.

EXTERNAL LEAK DETECTION SYSTEMS

External leak detection methods operate by monitoring the environment around a tank for evidence that product has been released. The regulation specifically includes two external methods, groundwater monitoring and vapor monitoring. Within each of these categories a wide variety of detector technologies is available.

Liquid-phase product detectors are generally deployed in groundwater wells surrounding the tank field to detect petroleum floating on the water table surface. The regulation requires these systems to detect 1/8 inch (0.32 cm) of the stored product when it is floating within a well. Groundwater monitoring must meet additional, site specific regulatory requirements related to water table depth, soil permeability and product characteristics.

Vapor-phase detectors are generally placed in vadose zone wells surrounding the tank field to detect the presence of (or an increase in) product vapors in the soil gas. Alternatively, soil vapor samples may be pumped to a centrally located control box that contains a sensor, or the sensor may be part of a portable unit used to sample a well. The regulation requires that vapor detectors be able to detect a significant increase in vapor concentration of the stored product above background levels. Vapor monitoring must meet additional site specific regulatory requirements regarding well location, soil permeability and product volatility.

For external monitors it is difficult to relate the measured quantity (e.g., vapor concentration) to a leak rate. The relationship depends on a number of site-specific and time-dependent factors including: soil permeability, soil moisture, temperature, groundwater level, groundwater gradient, product properties, and a number of others. Consequently, the test procedures for external methods determine the detector's ability to sense a specified quantity of product outside the tank without attempting to relate that quantity to a leak rate from the tank.

In general, the test procedures for external detection systems require that the sensor be exposed to test substances in varying amounts to determine its ability to reliably detect petroleum components at different levels. Qualitative detectors (those that merely provide a yes/no indication of whether contamination exceeds a preset threshold) are tested somewhat differently from quantitative detectors (i.e., those that measure the concentration of contaminants). The evaluation of quantitative detectors produces estimates of accuracy, precision, bias, response time, specificity and lower detection limit. The procedure for qualitative detectors provides similar information except that precision and bias are not applicable to these systems.

Test chambers must be constructed for each of the procedures. For liquid-phase detectors the test chamber consists of a machined cylinder with an accurately known cross-sectional area. The probe is suspended within the cylinder to simulate its placement in a well casing. For vapor-phase detectors the test chamber is a gas-tight vessel that is inert to the test gases. Test chamber volume is kept as small as possible without interfering with the detector's operation. In both procedures the dimensions of the test chamber are dictated by the size and shape of the detector probe.

Liquid-phase detectors are tested by placing water and a defined thickness of test product in the test cylinder, inserting the probe and recording the detector's output for 24 hours or until a steady high-level reading is obtained. Water temperature is also measured during testing. Actual product thickness is calculated from the cross-sectional area of the probe and the cylinder. The probe and the cylinder are cleaned between each test run. Two test products are used in the majority of the evaluation, commercial unleaded gasoline and a synthetic gasoline composed of 11 representative constituents of gasoline in fixed proportions. The synthetic gasoline is intended to eliminate performance variations caused by differences in composition and additives among brands of commercial gasoline. The test procedures are summarized in Table 1a.

Vapor-phase detectors are tested by running a 0.2 L/min stream of a defined concentration of test gas through the test chamber and recording the detector's output for 24 hours or until a steady high-level reading is obtained. Temperature and pressure within the test chamber are also recorded during testing. The test chamber is purged with ultrahigh-purity air between each test run. Two test gases are used in the majority of the evaluation, benzene and 2-methylbutane,

TABLE 1a--Summary of test procedures for liquid-phase detectors

Parameters Evaluated	Test products	Test product thickness	Total number of tests
accuracy*, precision**, bias**, response time	commercial gasoline synthetic gasoline	0.04, 0.32, 0.64 cm	30
specificity	commercial gasoline diesel fuel n-hexane Jet-A jet fuel synthetic gasoline toluene xylene(s)	1.27 cm	7
lower detection limit	commercial gasoline synthetic gasoline	0.01, 0.02, 0.04, 0.08, 0.16, 0.32, 0.64, 1.27 cm	14 (24)***

TABLE 1b--Summary of test procedures for vapor-phase detectors

Parameters Evaluated	Test products	Test product concentrations	Total number of tests
accuracy*, precision**, bias**, response time	2-methylbutane benzene	50, 250, 500, 1000 ppmv	40
specificity	benzene n-butane n-hexane isobutane 2-methylpentane 3-methylpentane toluene	500 ppmv	7
lower detection limit	2-methylbutane benzene	1.25, 2.5, 5, 12.5, 25, 50, 125, 250, 500, 1000 ppmv	14 (24)***

* Relative accuracy is computed for qualitative detectors
** Not applicable to qualitative detectors
*** Minimum number of tests for qualitative detectors within the test product thickness range

representing the aromatic and light alkane components of gasoline, respectively. Components are used rather than gasoline vapor because at the higher test concentrations the heavier constituents will condense making it difficult to control test gas concentrations and possibly damaging test equipment. Use of the pure components allows evaluating organizations to obtain gas cylinders at the correct concentrations from specialty gas vendors. The test procedures are summarized in Table 1b.

INTERNAL LEAK DETECTION SYSTEMS

Internal leak detection methods operate by sensing changes inside a tank that could indicate the presence of a leak. The regulation specifically includes several internal methods, tank tightness testing, manual tank gauging and automatic tank gauging. The majority of internal systems measure changes in the volume of product present in the tank over time. These methods are referred to as volumetric. Some internal methods do not measure volume and are referred to as non-volumetric. These methods have a variety of operating principles.

Volumetric tank tightness testing methods are generally performed by a service company under contract to the tank owner. The crew temporarily installs sensors in the tank that monitor level and temperature over the course of the test period, normally several hours. After correcting for volume changes due to temperature, the leak rate for the tank is calculated from the change in level during the test using an experimentally determined level to volume conversion factor.

Non-volumetric tank tightness testing methods attempt to measure product loss through indirect means. Two common forms of non-volumetric testing are acoustic methods (which listen for sounds in the tank associated with leakage) and tracer methods (which place a tracer material in the tank and monitor for its presence outside the tank). Unlike volumetric methods, most non-volumetric methods cannot determine the size of a leak. They merely report whether the tank is tight or not.

The operation of an automatic tank gauging system (ATGS) is typically similar to that of a volumetric tank tightness test except that the ATGS is permanently installed at the site. ATGS generally have both a daily inventory tracking function and a periodic precision test mode. Tank owners can begin a test by programming the equipment console.

Statistical inventory reconciliation vendors generally analyze a month or more of daily tank sticking records using proprietary software packages to look for consistent trends of product loss. In this method the amount and frequency of the data are supposed to offset its low quality. This method is not included in EPA's regulations, but vendors have claimed that the method can meet the performance standard for other methods. EPA developed the test procedure for this method to evaluate this claim.

For internal systems relating the measured quantity (e.g., change in product level) to leak rate is more straightforward than it is for external systems. Consequently, the performance of these methods is defined in terms of their ability to detect a specified leak rate with a specified reliability (probability of detection and probability of false alarm). A method's probability of detection for a certain leak rate represents the fraction of tanks the method would correctly declare to be leaking in a large sample of tanks leaking at that rate. The probability of detection (P_d) varies depending on the leak rate, increasing for larger leak rates. The probability of false alarm (P_{fa}) indicates the fraction of non-leaking tanks a method will mistakenly declare to be leaking. A method with a high P_d and a low P_{fa} can reliably differentiate between signal (a leak) and noise (other variables that may mask a leak). The regulation requires all internal methods to detect the leak rate specified for that method (e.g., 0.38 L/hr for tank tightness tests) with a P_d of 0.95 and a P_{fa} of 0.05 by December 23, 1990. A complete description of these terms is provided in reference [8].

The general evaluation approach for internal methods requires using the method to conduct a large number of tests of non-leaking tanks, some of them with simulated leaks. In most cases these tests are conducted on just one or two tanks at an evaluation facility. The test conditions are selected to subject the systems to typical interfering factors (noise). The differences between the actual (simulated) tank condition and that measured by the leak detection method form the basis for the evaluation.

Volumetric and non-volumetric tank tightness testing methods and automatic tank gauging systems are tested following similar procedures. Each evaluation procedure calls for using the system to repeatedly test the same non-leaking tank. The system is installed and operated following the manufacturer's normal test procedure. Key elements of this procedure are reported on a form included in each procedure. Leaks of specified sizes are simulated in a random order and important variables, such as the temperature of delivered product, are controlled to cover the range of conditions typically encountered in the field. The volumetric tank testing and ATGS protocol require at least 24 tests to be conducted. A minimum of 42 tests must be performed to evaluate non-volumetric methods.

When the UST is in contact with a high water table, water may flow into the tank instead of product flowing out. Detecting water in the tank is employed as a second mode of leak detection by many ATGS and several non-volumetric methods. Water intrusion is detected using a water sensor located in the bottom of the tank. The water sensors are tested to determine the minimum water level that they can detect and the smallest change in level that they can resolve. Water sensor tests are conducted in a test chamber similar to that used to test liquid-phase detectors.

An alternative evaluation approach is presented for automatic tank gauging systems that allows most of the test data to be collected from non-leaking, normally operating UST systems. Use of the operational tanks limits the need for controlling noise sources since these will be automatically included if enough tests are conducted. The alternative approach requires approximately 100 tight tank tests and 10 simulated leak tests as described above. The protocol suggests that the tight tank tests be conducted on a minimum of 10 tanks from at least 5 different sites.

Statistical inventory reconciliation methods are tested using tank sticking data obtained from non-leaking, operational UST sites. The evaluating organization gathers this data and manipulates some of it to introduce simulated leaks. If the method being evaluated provides qualitative results, data must be collected for a total of 80 tanks; quantitative methods require data from only 32 tanks. The length of time over which data is collected is determined by the amount of data required by the particular statistical inventory method. The data are provided to the vendor in a "blind format" for analysis according to the normal operating procedure of the method. For any necessary interactions the evaluating organization acts as the tank owner, receiving the results of the inventory analysis when it is completed. Results are obtained by comparing the vendor's conclusion about the leak status of each tank with the actual (simulated) condition.

PIPELINE LEAK DETECTION SYSTEMS

EPA's regulations allow both external and internal methods of leak detection to be used for piping. External methods of leak detection for piping are similar to those described above for external tank methods and are covered by the same test procedures. Internal piping methods generally monitor the pipeline for pressure, volume or flow-rate changes. They may be either permanently installed systems or may be line tightness tests conducted by a contractor on a less frequent basis.

The performance standards for pipeline leak detection systems, like those for internal tank systems, are defined according to a detectable leak rate, a probability of detection and a probability of false alarm. The requirements differ depending on whether the piping is pressurized (i.e., it has a submersible pump) or is a suction line. Since pressurized piping can cause catastrophic releases, the regulation requires it to be checked hourly for large leaks (11.4 L/h at 10 psi) and also checked for small leaks either (1) annually (0.38 L/h at 150% of operating pressure) (2) monthly (0.76 L/h at operating pressure) or (3) monthly with an external method. Suction piping does not pose the same hazards and only needs to be checked for small leaks infrequently. The test procedure covers internal leak detection methods for both pressurized and suction piping systems.

The evaluation approach for pipeline systems is similar to the procedures for internal tank methods. It requires collecting data on system performance in non-leaking pipelines under a wide range of test conditions. In addition, a limited amount of data is collected for some pipelines with simulated leaks. For each test run the system is installed and operated according to the manufacturer's instructions to determine if the line is leaking. The system's results are then compared to the actual condition of the line to determine its performance level.

The most important noise source for internal pipeline systems is the change in product temperature produced when product in the tank is pumped through the pipeline. A thorough evaluation will include a range of typical temperature conditions to test the system's ability to deal with this interference. The test procedure provides five options for gathering the required data; these options differ primarily in how the required temperature conditions are generated and how much data is needed. Manufacturers may select the procedure which is most convenient for them. Each option is listed below and then described further in the following paragraphs.

1. Collect data at a special pipeline test facility.
2. Collect data at one or more instrumented facilities.
3. Collect data over a 6- to 12-month period at 5 or more operational UST facilities.
4. Collect data over a 6- to 12-month period at 10 or more operational UST facilities.
5. Develop the data from an experimentally validated computer simulation.

Options 1 and 2 represent ways to control the environmental conditions and pipeline parameters to ensure an adequate evaluation. These options require the smallest amount of data and are the quickest to complete, but they involve the most labor-intensive data collection effort. The evaluating organization must either generate the required temperature conditions (option 1) or check off a matrix to verify that all of the necessary temperature conditions were present (option 2).

Options 3 and 4 require less control of conditions by the evaluating organization, but require more data and take longer to complete. In this approach the required temperature conditions are generated by normal seasonal temperature variations, and the required range of pipeline characteristics are obtained by using a number of different UST sites.

Option 5 is an approach similar to the one taken in reference [8] for tank testing methods. It is the most extensive of the evaluation options and few manufacturers are expected to choose this approach.

PRELIMINARY TEST RESULTS

In order to establish that the test procedures were both practical and applicable to a wide range of leak detection systems, EPA used draft versions of the test procedures to conduct evaluations of several commercial systems in 1988 and 1989. The manufacturers volunteered to have their systems evaluated on the condition of anonymity. The results of these early evaluations are presented as an example of how typical leak detection systems available at that time performed when evaluated as described above.

Results for Internal Methods

A total of four internal methods were tested including two volumetric tank testing systems, a non-volumetric system, and an automatic tank gauging system. The evaluation results are summarized in Table 2. Note that none of the four methods were able to achieve the levels of reliability required by EPA regulations (P_d=0.95 and P_{fa}=0.05). It is important to note, however, that simple procedural changes are often capable of dramatically improving the performance of internal methods [8]. A secondary goal of the test procedures was to motivate vendors to make the changes needed to improve the performance of their methods.

In addition to their utility in verifying regulatory compliance, the reported probabilities of detection and false alarm can also be used to assess the costs and benefits of using a particular method. A high false alarm rate, such as that for Vendor D, translates into costs for the tank owner in the form of unnecessary release investigation or, in the most extreme case, digging up a tank that is sound. Since this sort of error is often discovered by the tank owner, successful commercial systems will generally have low false alarm rates, such as those for Vendors B and C. On the other hand, a method with a low probability of detection can create long-term costs for owners by allowing a leaking tank to stay in the ground, necessitating a larger cleanup at a later date. This sort of error is harder for an owner to prove because a long time may elapse between a passed leak detection test and the discovery of a leak by other means. In this situation the vendor can always claim the leak began after the test was conducted. Consequently, it is common for commercial systems to have relatively low probabilities of detection, such as those reported for Vendors A, B and C.

Results for External Methods

Evaluation results for external methods are more difficult to interpret than those for internal methods, largely because the regulatory requirements on these systems are not as specific or quantitative. However, the results have great utility for tank owners or consultants selecting a method for a specific application.

As shown in Table 3, all six of the liquid-phase detectors tested were able to detect 0.32 cm (1/8 inch) of simulated gasoline in the test cell as required by the regulations. All of the methods except that of Vendor F exceeded this requirement and could detect 0.04 cm (1/64 inch)

TABLE 2--Preliminary test results
for internal leak detection methods

Vendor	P(D)	P(FA)	Leak Rate (gph)
Vendor A	63	17	0.1
Vendor B	62	4	0.2
Vendor C	40	3	0.1
Vendor D	94	46	0.1

TABLE 3--Preliminary test results
for liquid-phase detectors

Vendor	Lowest detectable Thickness (cm)	Detection time at 0.32 cm (min)
Vendor A	0.04	0.3
Vendor B	0.04	1.3
Vendor C	0.04	<0.02
Vendor D	0.04	0.2
Vendor E	0.04	64
Vendor F	0.32	0.1

Liquid-phase specificity tests
All detectors responded to
0.32 cm of product within:

Product	Detection times
n-hexane	<1 sec - 17.5 hr
Xylene	<1sec - 21.5 hr
Toluene	<1 sec - 1.5 hr
Simulated Gasoline	<1 sec - 1 hr
Commercial gasolin	<1 sec - 1 hr
Diesel Fuel	<1 sec - 26 hr
Jet-A jet fuel	< 1 sec - 16.5 hr

of product in the test cell. Most of the detectors responded to the presence of product within a minute, and only the method of Vendor E required more than an hour to respond. In the specificity testing, all six systems successfully detected 0.32 cm of the seven different test compounds within 26 hours. The reported variations in response time are not considered significant since days or weeks will likely elapse before enough product is released from a leaking tank to cause a detectable product layer to form in the wells where the detectors are located. For details on the time required for released product to travel to liquid or vapor wells around a tank field, refer to the paper "Analysis of UST Leak Vapor Diffusion and Liquid Build-up" by R.P. Schreiber and M.S. Rosenberg elsewhere in this volume.

The results for the preliminary evaluations of vapor-phase systems are shown in Tables 4 and 5. The six vapor-phase systems tested are separated based on whether they provide a qualitative or a quantitative output. All of the tested systems except that of Vendor B rapidly responded to the presence of 500 ppmv of benzene or isopentane. As for liquid-phase detectors, the variation in response time between the fastest detector (Vendor F, 0.25 min) and the slowest (Vendor C, 42 min) is not particularly significant in the choice of a system because the slowest detection step will be transport of the vapors from the point of the leak to the sensor location.

Each of the quantitative systems exhibited substantial bias in the measurement of one or both of the tested components. This level of bias does not necessarily represent a problem since each of the systems may be calibrated to a different type of gas, and the response would be expected to be exact (zero bias) only if the sensor were tested with the calibration gas at a concentration within the calibration range. If accurate concentration readings are needed for a specific application, however, the method of Vendor E, which greatly overestimates the actual concentration, would probably not be suitable. Perhaps the most important result of the evaluation is the precision measurement, which can be used to determine the smallest concentration change that a system can reliably discern. A value of precision near zero indicates a system that will be extremely reliable in detecting a significant increase in vapor concentrations above background levels, as required by the regulations.

Table 5 shows the results of the specificity testing. None of the three qualitative detectors (Vendors A-C) were able to detect any of the components tested at 250 ppmv. In the final test procedures, the specificity testing is conducted at 500 ppmv rather than the 250 ppmv of the preliminary testing. The different response characteristics of each quantitative detector can be clearly seen from these results. Vendor G's system has excellent response to the light alkanes tested, but is essentially unresponsive to aromatic components. These characteristics make it an excellent sensor for gasoline UST applications, since gasoline vapor is dominated by these fractions. The specificity of Vendor F's system is almost a perfect complement, providing accurate readings of aromatic concentrations, while being essentially unresponsive to alkane fractions. Vendor E's system responds to the entire range of components tested, but is not as accurate as those of Vendors F and G. Depending on the particular application, any one or a combination of these systems might be most appropriate.

TABLE 4--Preliminary test results
Vapor-phase detectors (500 ppm)

Vendor	Test Gas	Bias (percent)	Precision (percent)	Detection time (minutes)
Qualitative Methods				
Vendor B	Benzene	N/A	N/A	ND
	Isopentane	N/A	N/A	ND
Vendor C	Benzene	N/A	N/A	42
	Isopentane	N/A	N/A	42
Quantitative Methods				
Vendor D	Benzene	-56	8	8
	Isopentane	24	3	4
Vendor E	Benzene	150	0	0.5
	Isopentane	321	43	0.65
Vendor F	Benzene	-17	4	0.25
	Isopentane	-96	25	0.3
Vendor G	Benzene	-98	5	0.6
	Isopentane	-18	2	0.3

ND = Method did not detect compound
N/A = Measurement was not applicable to vendor's method

TABLE 5--Preliminary test results
Vapor-phase detector specificity (250 ppm)

Vendor	n-butane	2-methyl-pentane	n-hexane	Benzene	Toluene
Vendor A	ND	ND	ND	ND	ND
Vendor B	ND	ND	ND	ND	ND
Vendor C	ND	ND	ND	ND	ND
Vendor D	11	65	61	42	17
Vendor E	100	200	200	100	150
Vendor F	1	12	10	100	102
Vendor G	68	92	96	2	14

ND = Method did not detect the compound

CONCLUSION

The standard test procedures described in this paper should improve the quality and consistency of information available to everyone concerned with underground storage tank leak detection. Using the results from these procedures, tank owners will be able to readily determine if a method meets the regulations and their specific needs before purchasing it. State and local regulators who lack the engineering expertise to evaluate methods on their own will be able to approve methods with more confidence that they are protecting the public. Finally, reputable leak detection manufacturers will be able to prove that their method lives up to the performance claims they have made for it, and may have an easier time of getting the method approved on a nationwide basis.

REFERENCES

[1] U.S. EPA, Solid Waste and Emergency Response/Research and Development. (1990) <u>Standard Test Procedures for Evaluating Leak Detection Methods: Liquid-Phase Out-of-Tank Product Detectors</u>. (EPA/530/UST-90/009)

[2] U.S. EPA, Solid Waste and Emergency Response/Research and Development. (1990) <u>Standard Test Procedures for Evaluating Leak Detection Methods: Vapor-Phase Out-of-Tank Product Detectors</u>. (EPA/530/UST-90/008)

[3] U.S. EPA, Solid Waste and Emergency Response/Research and Development. (1990) <u>Standard Test Procedures for Evaluating Leak Detection Methods: Volumetric Tank Tightness Testing Methods</u>. (EPA/530/UST-90/004)

[4] U.S. EPA, Solid Waste and Emergency Response/Research and Development. (1990) <u>Standard Test Procedures for Evaluating Leak Detection Methods: Non-volumetric Tank Tightness Testing Methods</u>. (EPA/530/UST-90/005)

[5] U.S. EPA, Solid Waste and Emergency Response/Research and Development. (1990) <u>Standard Test Procedures for Evaluating Leak Detection Methods: Automatic Tank Gauging Systems</u>. (EPA/530/UST-90/006)

[6] U.S. EPA, Solid Waste and Emergency Response/Research and Development. (1990) <u>Standard Test Procedures for Evaluating Leak Detection Methods: Statistical Inventory Reconciliation Methods</u>. (EPA/530/UST-90/007)

[7] U.S. EPA, Solid Waste and Emergency Response/Research and Development. (1990) <u>Standard Test Procedures for Evaluating Leak Detection Methods: Pipeline Leak Detection Systems</u>. (EPA/530/UST-90/010)

[8] U.S. EPA, Research and Development. (1988) <u>Evaluation of Volumetric Leak Detection Methods for Underground Fuel Storage Tanks, Vol. I</u>. (EPA/600/2-88/068a)

William D. Glauz,[1] Jairus D. Flora,[1] G. Joe Hennon[1]

EVALUATION OF PIPELINE LEAK DETECTION SYSTEMS

REFERENCE: Glauz, W. D., Flora, J. D., and Hennon, G. J., "Evaluation of Pipeline Leak Detection Systems," Leak Detection for Underground Storage Tanks, ASTM STP 1161, Philip B. Durgin and Thomas M. Young, Eds., American Society for Testing and Materials, Philadelphia, 1993.

ABSTRACT: Leaking underground storage tank systems present an environmental concern and a potential health hazard. It is well known that leaks in the piping associated with these systems account for a sizeable fraction of the leaks. EPA has established performance standards for pipeline leak detection systems, and published a document presenting test protocols for evaluating these systems against the standards. This paper discusses a number of facets and important features of evaluating such systems, and presents results from tests of several systems. The importance of temperature differences between the ground and the product in the line is shown both in theory and with test data. The impact of the amount of soil moisture present is addressed, along with the effect of frozen soil. These features are addressed both for line tightness test systems, which must detect leaks of 0.10 gal/h (0.38 L/h) at 150% of normal line pressure, or 0.20 gal/h (0.76 L/h) at normal line pressure, and for automatic line leak detectors that must detect leaks of 3 gal/h (11 L/h) at 10 psi (69 kPa) within an hour of the occurrence of the leak. This paper also addresses some statistical aspects of the evaluation of these systems. Reasons for keeping the evaluation process "blind" to the evaluated company are given, along with methods for assuring that the tests are blind. Most importantly, a test procedure is presented for evaluating systems that report a flow rate (not just a pass/fail decision) that is much more efficient than the procedure presented in the EPA protocol, and is just as stringent.

KEYWORDS: pipeline, leak detection, temperature difference, soil moisture, line tests, bulk modulus, statistical evaluation

INTRODUCTION

The Environmental Protection Agency (EPA) promulgated regulations (40 CFR Part 280) providing technical standards and corrective action requirements for owners and operators of underground storage tanks. These were published on September 22, 1988, and the various elements of the regulations become effective on a phased time schedule. Included in the definition of an underground storage tank are the underground pipes connected to them that typically lead from the tank to the points of fuel distribution or delivery. Because these pipelines are pressurized in most tank installations, and are usually field-installed, leaks in "underground storage tanks" are often leaks from the piping. Some tank

[1] Principal Advisor, Senior Advisor, and Principal Chemist, respectively, Midwest Research Institute, 425 Volker Boulevard, Kansas City, MO 64110

installations utilize suction piping, which also falls under the regulations, but will not be dealt with explicitly here. Also, this paper does not consider cross country transmission lines, natural gas distribution lines, etc.

As a means of reducing the possibility of leaking regulated substances to the environment, the regulations identify three types of tests for discovering pipeline leaks:

1. An annual test, using a system or method that can detect a leak of 0.1 gal/h (0.38 L/h) at 1.5 times the pipeline's normal operating pressure, "with a probability of detection of [at least] 0.95 and a probability of a false alarm of [not more than] 0.05".

2. A monthly test, using a system or method that can detect a leak of 0.2 gal/h (0.76 L/h) at the pipeline's normal operating pressure, "with a probability of detection of [at least] 0.95 and a probability of a false alarm of [not more than] 0.05".

3. An "hourly test", using a system or method that can detect a leak of 3.0 gal/h (11 L/h) at 10 psi (69 kPa) "with a probability of detection of [at least] 0.95 and a probability of a false alarm of [not more than] 0.05" within 1 h.

EPA published a series of 7 procedures or protocols for evaluating methods or systems, to establish that they meet the EPA criteria for leak detection. One of these protocols [1], hereafter referred to as the line test protocol, for brevity, deals with the evaluation of systems to be used for any of the above three types of tests. This paper presents a number of findings of the authors as a result of conducting evaluations of pipeline leak detection systems following the procedures presented in the line test protocol.

EFFECTS OF CIRCULATION ON GROUND TEMPERATURE

The line test protocol requires that the system to be evaluated be challenged with a number of tests on a pipeline under both a tight condition and with a simulated leak. A major factor that can interfere with the system's ability to correctly determine whether the line is tight or leaking is the temperature differential between the product in the pipeline and the ground.

The protocol specifies the range of temperature differentials to be used, and further requires that the product be circulated through the pipeline for at least an hour prior to a test. This requirement is based on the fact that if the product in the pipe is warmer or cooler than the ground, circulation will tend to warm or cool the ground near the pipe. When the circulation is stopped, the product will take some time to come to equilibrium with the ground, which can interfere with the test. This would not occur if the product had not been circulated. Further, such circulation simulates the actual operation of a service station which periodically and frequently delivers product to the pumps.

Several figures, taken from test data, are presented that illustrate the warming and cooling phenomena. Fig. 1 shows the change in ground temperature with time when product (diesel fuel) at 108°F (42°C) was circulated through the ground at an initial uniform temperature of 83°F (33°C). The product was circulated in a nominal 2 in. (5 cm) diameter fiberglass pipeline. Temperature measurements were taken at distances of 2, 4, and 12 in. (5, 10, and 30 cm), respectively, from the pipeline, in accordance with the line test protocol. Note that the temperature of the ground closest to the pipe, T_2, increased about 6°F (3°C), whereas it was virtually unaffected at 12 in. (30 cm).

Fig. 2 illustrates the inverse situation. Here the ground was initially a little cooler close to the pipe than at 12 in. (30 cm) because of prior tests, and then product at 57°F (14°C) was circulated for an hour. T_2 dropped about 7°F (4°C) during this period.

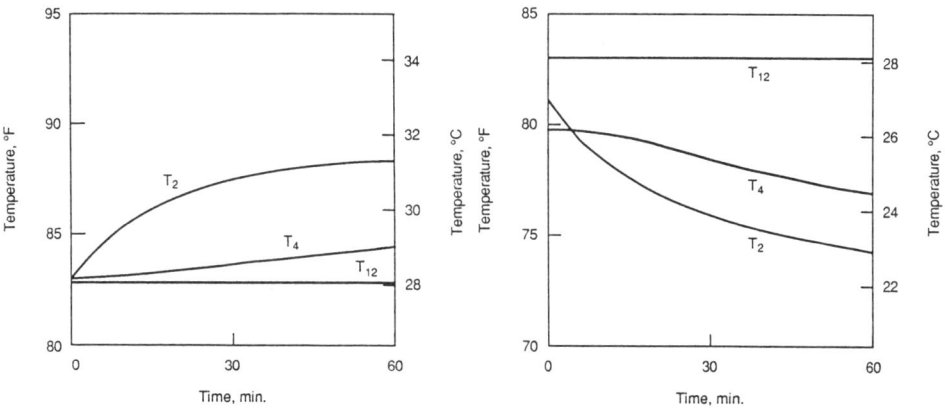

Fig. 1--Ground Temperature Change Circulating Warm Product.

Fig 2--Ground Temperature Change Circulating Cool Product.

Fig. 3 illustrates a natural phenomenon which might interfere with an actual test. In this case a rain shower occurred during the testing period, warming up the ground quite uniformly.

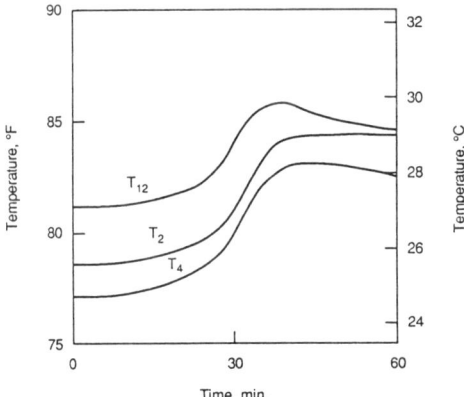

Fig. 3--Effect of Rain on Ground Temperature During Testing.

EFFECTS OF TEMPERATURE DIFFERENTIALS ON LEAK DETECTION

General

As the product in the pipe is cooled or warmed by heat transfer to or from the ground, its volume will tend to shrink or expand in accordance with the relationship:

154   LEAK DETECTION FOR UNDERGROUND STORAGE TANKS

$$\Delta V/V = \gamma \, \Delta T \qquad (1)$$

where γ is the coefficient of volume expansion. (The pipe, itself, will also contract or expand, but that change will not be dealt with here.) The effects of temperature changes are potentially very important with annual or monthly test methods, although the hourly tests are so rapid that temperature changes are usually (but not always) insignificant.

As a numerical example, diesel fuel has a coefficient of expansion of approximately 0.0005 /°F (0.0009 /°C). A nominal 2 in. (5 cm) diameter pipeline that is 133 ft (40.5 m) long contains a volume, V, of approximately 5000 in.^{3} (82 000 cm^3). If the product in the pipe cools by 5°F/h (3°C), the rate of volume change is, using Eq 1, 12.5 in.^{3}/h (205 cm^3/h), equivalent to an apparent leak rate of 0.054 gal/h (0.20 L/h). This is roughly the detection threshold of most annual line leak detection systems that are designed to detect a leak rate of 0.10 gal/h (0.38 L/h). Thus, such a system, in order to reliably detect such small leaks, must either wait until the rate of temperature decrease is substantially less than this, or else have a means to distinguish between volume changes due to temperature declines and leaks. Conversely, if the temperature is increasing, there would tend to be a volume expansion in accordance with Eq 1. This expansion could mask a volume decrease due to a leak, so must also be accounted for.

Several sketches illustrative of tests of a system indicate these features. During the tests the tank's submersible pump was cycled periodically, bringing the line to normal operating pressure. The system then made a rapid, proprietary analysis of the piping response. Next the pump was shut off, the pressure allowed to drop to a lower level, and the underground piping isolated by a valve closure. The piping then remained dormant until it was time for the next cycle.

Fig. 4 illustrates a series of cycles in which warm product was circulated prior to the test. Note the pressure drop during each cycle as the cooling product contracted. If, in this test, the line was known to be tight, one can be confident that the pressure drops were not caused by loss of product. Fig. 5 illustrates a similar test with a cooling product, but also with a simulated leak. The challenge to the system is to distinguish cases like Fig. 4 from those like Fig. 5.

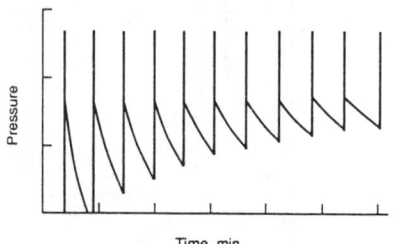

Fig 4.--Tight Line With Warm Product.

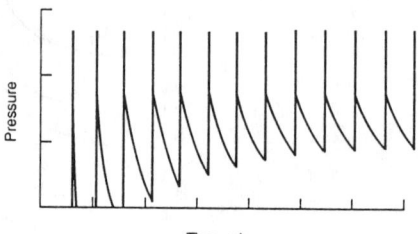

Fig. 5--Leaking Line With Warm Product.

An example with warming (expanding) product is given in Fig. 6. If one only examined the first few cycles, one would be tempted to decide the line was tight. However, a simulated leak is more apparent after an hour or two.

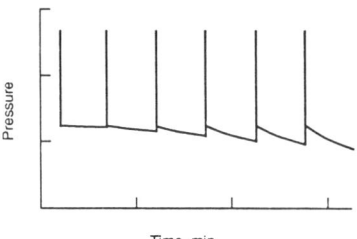

Fig. 6--Leaking Line with Cool Product.

Effect of Soil Moisture

A factor which can affect the rate of temperature change through the ground surrounding the pipeline, and thus the rate of expansion or contraction of the product inside the pipe, is the amount of moisture in the soil or back fill. Dry soil is a poor conductor of heat whereas water or wet soil is a good conductor. Thus, wet soil will transfer heat to or from the pipe more rapidly than dry soil.

The temperature at a point in the soil, at the surface of the pipe for example, can be expressed as:

$$T - T_\infty = (T_0 - T_\infty) \exp(-xt) \qquad (2)$$

where T is the temperature at time, t, T_0 is the initial temperature, T_∞ is the temperature of the soil far from the pipe, t is the time, and x is a coefficient dependent on the heat transfer characteristics of the soil. The more moisture in the soil, the larger is x. The wetter the soil (larger x) the more rapidly the temperature approaches its equilibrium value.

Fig. 7 illustrates the effect of soil moisture on the heat transfer through the soil. This illustration shows warm product cooling to the soil temperature. What is the impact of the soil moisture on the leak detection system? The critical issue is the length of time it takes for the rate of change of temperature to reach an acceptably low level. It is not obvious from the curves in Fig. 7 as to which soil condition is "best" from this viewpoint. What needs to be examined is the time rate of change of temperature (the time derivative of T). An illustration of the derivative is shown as Fig. 8. Note that the "wet" soil has an initially higher derivative, but it becomes smaller more rapidly than the "moist" soil.

Also shown in Fig. 8 are two temperature derivative thresholds, ϵ_1 and ϵ_2. In this example, if the upper threshold is acceptable, then the moist soil would enable a leak test sooner than the wet soil. However, if the lower threshold prevails the reverse is true. The heat transfer rate of soils, represented here by x, can easily vary by two or more orders of magnitude (the range illustrated in Fig. 7), depending on the amount of moisture present. The effect of this on evaluation of leak detection methods can obviously be significant. The means of dealing with this effect have not yet been quantified, and are not dealt with in the line test protocol. More research is required on this matter.

156　LEAK DETECTION FOR UNDERGROUND STORAGE TANKS

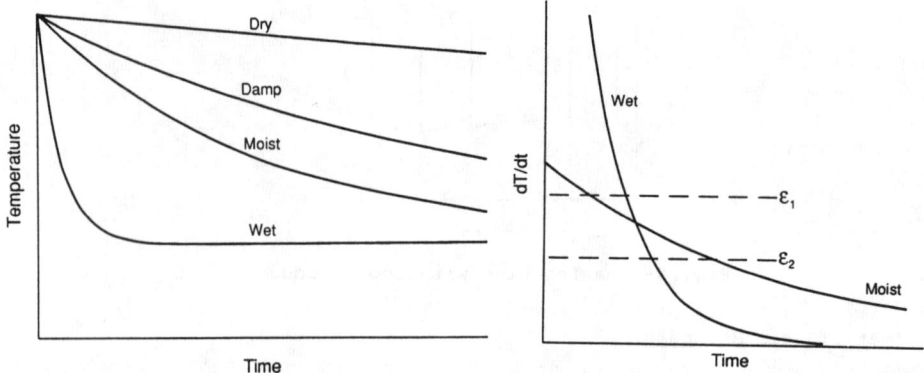

Fig. 7--Effect of Soil Moisture on Temperature.

Fig. 8--Effect of Soil Moisture on Temperature Derivative.

Frozen Soil

If the moisture in the soil adjacent to the pipeline freezes or is frozen, the heat transfer characteristics change dramatically. Whereas moist or wet soil is a good conductor of heat, ice or frozen soil is not. Further, a relatively large amount of heat (the heat of fusion) is required to freeze water or moisture in soil, or to melt ice or frozen soil, and the phase change occurs with no change in temperature. Thus, for example, if the ground temperature is on the order of 40°F and product with a temperature 25°F colder is circulated, the moisture in the soil near the pipe may become frozen. Then, after circulation is stopped, the temperature of the product in the pipeline may look somewhat like the solid line in Fig. 9. It will warm up more gradually than if there were no ice (dashed line), due the poorer heat transfer of the ice. Then, it will stabilize at approximately 32°F until the adjacent soil melts, after which the product will resume its warming trend as the now moist soil conducts heat to the pipe. For this reason, we have been cautious in performing evaluations when the soil temperature approaches the freezing point, because a large negative temperature differential between product and ground in this case may not sufficiently challenge the leak detection method.

Another, and related, issue is performing a pipeline leak detection test when the soil around the pipe is frozen to begin with. In the northern states in the winter the frost line is often below the depth of the piping. The product in the tank, that may be mostly below the frost line, may be warmer than the soil (which, for purposes of this discussion, is taken as somewhat less than 32°F). The solid line in Fig. 10 illustrates how the temperature of the product in the pipeline might decrease with time during a line leak test, compared with its behavior in the absence of ice (dashed line). The temperature will decrease to about 32°F, melting some of the frozen moisture in the soil as it does so. It will then stay relatively constant until the soil refreezes, and then cool at a slower rate because of the poorer heat transfer properties of the frozen soil. The point to be made, however, is that this is another issue where more research is needed, and that is not dealt with by the line test protocol.

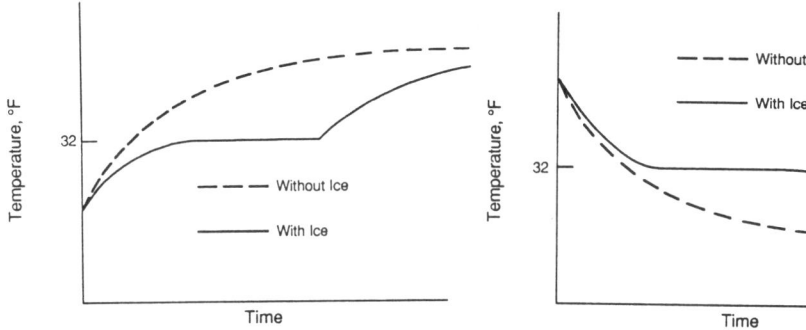

Fig. 9--Circulation of Very Cold Product.

Fig. 10--Circulation of Product in Frozen Soil.

BULK MODULUS

Definition

The bulk modulus, B, is a measure of the stiffness of the pipeline and its contents, and indicates how much the pipeline will expand under pressure. It is defined, mathematically, as:

$$B = \Delta P / (\Delta V/V_0) \qquad (3)$$

where V_0 is the initial volume of product in the pipeline, and ΔP and ΔV are the pressure and volume change, respectively. The value of B depends primarily on two phenomena. (Other factors affecting the bulk modulus are discussed later in this section.) First, the liquid in the pipeline, whether it be gasoline, diesel, or another product, is not incompressible (as is often assumed in beginning physics). The bulk modulus of the fuel alone is about 150 000 psi for gasoline, 250 000 psi for diesel (1000 MPa and 1700 MPa, respectively). Secondly, the pipeline material will expand under pressure, both in diameter and in length. The amount depends on the actual geometry of the pipeline and its material; fiberglass expands more than steel and thus has a lower B.

The line test protocol specifies that the bulk modulus of a test facility used to evaluate a line leak detection system be 25 000 ± 10 000 psi (172 ± 69 MPa). This means, using Eq 3, that an increase in pipeline pressure of 25 psi (172 kPa) will cause a pipeline with the nominal B to expand by 0.1 %, ($\Delta V \backslash V_0 = 0.001$) or by 5 in.^{3} (82 cm^3) if its initial volume is 5000 in.^{3} (82 000 cm^3).

Implications of the Bulk Modulus on Ability to Detect Leaks

Most leak detection methods used for pipelines use a variant of one of three fundamental ideas: observe the pressure drop with time when the pipeline is isolated from the pump and no product is added; isolate the system and measure the amount of product added after a period of time to bring the pressure back to a preset level; or measure the rate of product addition needed to maintain a constant pressure. The effect of B on each of these three types of methods is different. In the following discussion, it is assumed that if alternative test methods can each detect the EPA-mandated leak rate, the preferred method is that which can detect the leak the most quickly.

Consider, first, leak detection methods based on analyzing the drop in pressure over time that results from product leaking from the pipeline. The bulk modulus is important for such methods because, assuming for the moment that there is no ΔT between product and ground, the pressure will drop faster with stiffer piping than with softer piping. This is clear if we rewrite Eq 3 as:

$$dP/dt = B\ V_0\ dV/dt \qquad (4)$$

Thus, a given leak rate, dV/dt, should be detectable more quickly in piping systems with larger B than in systems with smaller B. For example, in a pipeline with the nominal B specified in the line test protocol with a leak rate of 0.1 gal/h (0.38 L/h) at 150% of normal operating pressure, the pressure will drop at a rate of approximately 2 psi per min (14 kPa per min). If the bulk modulus is twice as great, the pressure will drop twice as fast; if it is half as great, the pressure will drop half as fast. Thus, again neglecting temperature effects for the moment, these types of leak detection methods should be able to detect leaks more quickly with stiffer (higher B) piping, and require a longer test with softer (lower B) piping.

Methods that periodically measure product additions are affected differently by the bulk modulus. As the pressure drops in a pipeline the leak rate through a hole will decrease. Because the leak rate is not constant with time, any estimate of the leak rate by the method will have some error. If the method is quantitative (it reports a numerical value of a leak rate), its reported value will be low. If it is qualitative (it reports only a pass or fail decision), it may also err -- it must compare its estimated leak rate with a pass/fail threshold. These methods will do better on pipelines with a smaller B because the pressure will drop more slowly and the leak rate will be more nearly constant. Pipelines with high values of B will lose pressure more rapidly (the pressure may actually reach zero and the leak stop), making it more difficult for these methods to accurately measure leaks.

Another way to view these methods is to rewrite Eq 3 as:

$$\Delta V/V_0 = \Delta P/B \qquad (5)$$

For a given pressure drop, the volume change, is inversely proportional to B. If the precision of the method is measured in terms of the ΔV detectable, the method is more precise on piping systems with smaller B.

Finally, there are leak detection methods that keep the pressure constant and measure the amount of product that needs to be added, if any, to maintain the pressure. If the pressure is truly constant, then so is the geometric volume of the piping system. Thus, again in the absence of thermal effects, the amount of product added is precisely equal to the amount lost through leakage, regardless of the value of B.

Effect on Thermal Expansion/Contraction

As noted earlier, changes in the temperature of the product cause it to expand or contract. The bulk modulus of the pipeline will affect the way it responds to temperature changes. Combining Eqs 1 and 3 yields:

$$\Delta P = B\ \gamma\ \Delta T \qquad (6)$$

which is the pressure change in the pipeline, ΔP, corresponding to a temperature change, ΔT.

Using the nominal bulk modulus and the value of γ for diesel fuel as before, Eq 6 indicates that a 1°F (0.6°C) decrease in temperature will cause a 12.5 psi (86 kPa) drop in pressure. This could be mistaken for a leak by a method that observes pressure drops, unless it has a means of distinguishing temperature effects from leaks.

In the application of some test methods, the rate of change of pressure may be more meaningful than the magnitude of the change. Dividing each side of Eq 6 by Δt, an interval of time, the resulting equation can be used to examine the rate of change, or approximate time derivatives, of temperature and pressure. For example, again using diesel fuel and the nominal value of bulk modulus, a 5°F/hr (3°C/hr) rate of change of temperature corresponds to a 62.5 psi/h (430 kPa/h) rate of change of pressure. Stated differently, the pressure would change at the rate of 0.96 psi/min (7.2 kPa/min), or a 10 psi (69 kPa) change would require 9.6 min. If the pipeline had a value of B twice as large (was stiffer), it would require only 4.8 min for that amount of pressure change; it would take twice as long if B was half as great.

Thus, the bulk modulus affects the rate of pressure change resulting from temperature changes. However, as noted earlier the bulk modulus also affects the rate of pressure change resulting from a leak. Increasing the bulk modulus makes the piping respond faster to both temperature changes and leaks. In summary, the value of the bulk modulus has no effect on any method's ability to distinguish between a temperature effect and a leak.

Parameters Affecting Bulk Modulus

It was indicated earlier that the bulk modulus is different for different liquids (eg., diesel fuel and gasoline). It is also dependent on other parameters [2]. It is mildly dependent on pressure, which effect can be ignored at the modest pressures in underground storage tank piping, but which could be important in high pressure hydraulic piping. The bulk modulus for hydrocarbons decreases on the order of 0.1% per degree Fahrenheit (0.2% per degree Celsius) increase in temperature, or a maximum of 2 to 3% over the temperature differentials specified in the line test protocol. These are probably not of consequence.

Another parameter affecting bulk modulus is entrapped air. This effect, which is pressure-dependent, can be fairly large at the low pressures in the piping used with underground storage tanks [2]. We have observed this effect with diesel fuel during the one-hour circulation period in a test facility that dumps the circulated product back into the tank, with splashing and foaming. Apparent errors of 20 to 30 percent were observed.

Finally, the properties of the backfill could also influence the bulk modulus. A backfill material or its compaction, if it is sufficiently unyielding, could hinder the free expansion of the pipeline under increasing pressure, thus increasing its apparent bulk modulus. The magnitude of this effect is unknown.

BLIND TESTING

The line test protocol discusses various aspects of blind testing, but leaves it to the evaluator as to whether to keep the test conditions blind to the firm whose method is being evaluated. In the line test protocol and in our experience, there are two aspects of the testing to be considered in reaching a blind-test decision. These are the ΔT and the presence or absence of a simulated leak.

Temperature Differential

We have found it rather impractical to keep the temperature differential blind to the manufacturer. Especially at the more extreme differences such as ±25°F (±14°C), the very warm or very cold product (and, consequently, the pipeline) are immediately evident. Further, we have on occasion performed evaluations at the test facility of the manufacturer, using their heating/cooling equipment, in which case the manufacturer probably knows the difference precisely.

We do not believe that it is necessary, with the exception to be explained subsequently, that the temperature differential be kept blind. In fact, in the field the testing firm could, in principle, measure the temperature of the tank and the ground before making a test (some firms, in fact, do this). Systems that perform an hourly test are automated, so a priori knowledge of the temperature differential would not affect the test results. Similarly, automated systems that perform monthly or annual tests without human intervention, other than to start the test, would not be affected by such knowledge. However, a system whose logic attempts to account for the temperature differential, by using a longer stabilization time for example, is an exception. With this type of system the manufacturer should not be told the temperature differential unless the system makes its own measurements or otherwise has no way of making use of the evaluator's data.

Leak Condition

We strongly believe that the leak condition set by the evaluator for each test must be kept blind to the manufacturer in all cases to maintain authenticity of the tests. Our experience is that most manufacturers desire an honest evaluation, but there is always the human instinct to want to take advantage of all available data. Even with a totally automated system, if a test is done in the presence of the manufacturer, he is not told the test condition. (Obviously, if the test is done without the manufacturer present the point is moot.)

One complication to keeping the testing blind arises if the procedures in the line test protocol are followed exactly. If, at each temperature condition, two tests are run in sequence (as the line test protocol allows under some conditions) and one test is with a leak and one without, presumably the manufacturer would have two test results that he could compare. He might be able to make the pass/fail decision (if that is what his system does) based on a comparison of the two results, rather than looking at each of them independently.

Two steps can be taken to minimize this possible compromise to the test series. One step is to require the manufacturer to "call" the result of each test before the next is begun. We generally do this. The second step is to depart slightly from the test schedule in the line test protocol, and to insert some pairs in the schedule with each of the two tests being a pass (or a fail). This we believe to be acceptable as long as the total number of tight tests and leaking tests meets the specifications of the line test protocol and they cover the prescribed temperature range.

EVALUATION OF A QUANTITATIVE METHOD

Many pipeline leak detection systems are qualitative, in that the result of a test is a decision -- pass or fail. The line test protocol describes the statistical analysis to be used with the data obtained in such testing, which is basically to divide the number of correct (or incorrect) decisions by the number, n, of tests performed, not by n+1 as the line test protocol incorrectly indicates.

However, some methods produce a quantitative estimate or measure of the leak rate. That is, they report a calculated leak rate, which in an evaluation could be compared to the actual leak rate. The line test protocol also describes how such data should be analyzed. The procedure given is essentially equivalent to the qualitative analysis procedure; this equivalence is acknowledged in the line test protocol. The procedure described in the protocol looks only at the "tail" of the cumulative distribution of the test results, either near the 5% or 95% points of the distribution depending upon whether the probability of false alarm or the probability of detection is to be determined. Doing so is inefficient in a statistical sense, because the numerical results of all the other tests are not utilized. The net effect of this is to require more tests to be performed than are really required to obtain the same confidence in the results.

The procedure we have adopted uses a more efficient data analysis approach to take advantage of the additional information contained when one can compare the measured leak rates and the actually induced leak rates. The data analysis procedure is identical to that in the EPA protocol for evaluating volumetric leak detection devices for underground storage tanks [3]. The test procedure used calls for 32 tests to be conducted, eight at each of four different leak rates. (The volumetric tank test method requires only 24 tests.) The larger number we used with line tests assured that this design would be at least as stringent as that provided in the line test protocol.

The approximate efficiency of the data analysis procedure described in the line test protocol relative to the procedure we recommend is about 64%. That is, the data analysis procedure in the line test protocol uses only about 64% of the available information in the data. This means that an analysis procedure that uses all of the information in the data needs only 64% as many data points to achieve the same levels of confidence. (Statistically, the procedure in the line test protocol is equivalent to the sign test, while the procedure recommended here is a t-test. The asymptotic relative efficiency of the sign test to the t-test when the errors are normal is $2/\pi$ or about 64%.) Consequently, the recommended evaluation procedure is at least as stringent as that described in the line test protocol and more stringent than the volumetric tank test protocol.

REFERENCES

[1] Standard Test Procedures for Evaluating Leak Detection Methods: Pipeline Leak Detection Systems, EPA/530/UST-90/010, September, 1990.

[2] Technical Hydraulic Handbook, Lee Company, Westbrook, CT.

[3] Standard Test Procedures for Evaluating Leak Detection Methods: Volumetric Tank Tightness Testing Methods, EPA/530/UST-90/004, March 1990.

Shelda A. Sutton-Mendoza[1]

EXPEDITED ENFORCEMENT OF UST REGULATIONS IN NEW MEXICO

REFERENCE: Sutton-Mendoza, S. A., "Expedited Enforcement of UST Regulations in New Mexico," Leak Detection for Underground Storage Tanks, ASTM STP 1161, Philip B. Durgin and Thomas M. Young, Eds., American Society for Testing and Materials, Philadelphia, 1993.

ABSTRACT: New Mexico is the first state to implement an expedited enforcement program to enhance compliance with the Underground Storage Tank Regulations (USTR). UST field inspectors conduct inspections at UST facilities to ensure compliance with the USTR. If the inspector detects a violation, a field Notice of Violation is issued to the owner/operator and a penalty assessed. Prior to expedited enforcement 14% of the violations were corrected. Since implementing the program the Underground Storage Tank Bureau has conducted 592 inspections, issued 181 field Notices of Violation and 52% of those cited have complied. As a result of the enforcement program 85% of the facilities inspected are in compliance with USTR. This program has been successful in gaining compliance with USTR, specifically the release detection requirements.

KEYWORDS: release detection, underground storage tanks, enforcement, citations, New Mexico, Regulations, inspections

The New Mexico Underground Storage Tank Program is administered by the Underground Storage Tank Bureau, New Mexico Environment Department (NMED). The Underground Storage Tank Regulations were promulgated by the Environmental Improvement Board pursuant to the New Mexico

---

[1]Program Manager, Prevention Inspection Section, Underground Storage Tank Bureau, NM Environment Department, P.O. Box 26110, Santa Fe, NM 87502

Hazardous Waste Act, subsection 74-4-1 et seq NMSA 1978. The release detection, new tank and upgrade requirements track precisely with the federal UST Regulations. In September 1990 the Prevention/Inspection Section, Underground Storage Tank Bureau implemented a program to conduct inspections at UST facilities to determine compliance with the release detection requirements. The program utilized 13 field inspectors in 9 locations throughout the state. The highest priority of the Section was to conduct release detection compliance inspections at facilities with tanks installed prior to 1965 or age unknown located in areas where the depth to ground water is less than 100 feet. If the inspector noted a violation during his inspection he wrote a letter to the owner/operator informing him of the violation and seeking compliance with the Regulations. For serious violations a Notice of Violation was issued by the Bureau Chief. If an owner/operator fails to comply with the Notice of Violation, a Compliance Order may be issued which might include a civil penalty of $10,000 per tank per day of noncompliance or the Bureau may commence a civil action in state district court. However, the Bureau had very little success with this program and it was very lengthy and time consuming. In addition, enforcement actions involved an attorney which was very costly. Consequently, the Bureau sought to develop an effective enforcement program that would be more efficient, more timely, and less costly.

February 1, 1991, New Mexico implemented the first statewide expedited enforcement program. At this time, the priority of the Section was to conduct release detection compliance inspections at facilities with UST systems installed prior to 1970 and located in areas where depth to groundwater is less than 100 feet. If the inspector detects a violation of USTR while conducting these inspections, a field Notice of Violation ("citation") is immediately issued to the owner/operator of the UST system, pursuant to subsection 74-4-10 NMSA 1978. A prelitigation settlement penalty is assessed at the time of the inspection or sent by certified mail immediately following the inspection. Most violations, including release detection, carry a $100.00 per violation penalty. Violations concerning proper tank closure or change in service carry a $500.00 penalty. The owner/operator must pay the penalty, correct the violation and certify compliance within 30 days of the receipt of the citation. If the violation is not corrected and/or the penalty is not paid within 30 calendar days of the citation, the NMED may exercise its prosecutional discretion to issue an administrative compliance order, pursuant to subsection 74-4-12 NMSA 1978, or commence a civil action in state district court against the owner/operator. A Compliance Order may include a civil penalty not to exceed $10,000 per tank per day of noncompliance. Failure to comply with the Compliance Order may result in the assessment of civil penalties not to exceed $25,000 per tank per day of noncompliance with the order.

RESULTS

From September 1990 to January 31, 1991 field inspectors conducted 433 compliance inspections at UST facilities. The inspectors issued 86 violation letters and 12 of these or 14% complied (Table 1).

TABLE 1--Inspection data prior to expedited enforcement

Number of Compliance Inspections September, 1990 - January 31, 1991	433	...
Number of Violation letters issued August, 1990 - January 31, 1991	86	20%
Number of above violations corrected	12	14%

Beginning February 1, 1991 the USTB initiated the expedited enforcement (citation) program. The results of the program are presented in Table 2. From February 1, 1991 to October 31, 1991 the field inspectors conducted 592 inspections and issued 181 field Notices of Violation (citations). About 69% of the facilities inspected were found to be in compliance while 31% were not. Ninety-four or 52% of the 181 non-compliant facilities corrected the violations and were in compliance with the USTR. A total of 85% of facilities inspected are now in compliance with USTR. As a result of failure to correct violations and/or pay penalties, five matters have been referred to the Office of General Counsel for commencing judicial action and seeking the statutory maximum penalty of $10,000 per day per tank for each violation. One of these matters has been filed in District Court for injunctive relief and civil penalties.

TABLE 2--Expedited enforcement data February 1-October 31, 1991

Number of Compliance Inspections	592	...
Number of Facilities in Compliance	411	69%
Number of field Notice of Violation Issued	181	31%
Number of Informal Conferences	53	29%
Number of field Notice of Violation Corrected	94	52%
Number of Penalties Paid	82	45%
Number of Violations Corrected and Penalties Paid (if applicable)	84	46%
Total Number of Facilities in Compliance	505	85%
Total Amount of Penalties Paid	$14,900.00	

The 181 field Notices of Violation were issued for 398 violations of the UST Regulations. The violations fall into three main categories: failure to register, failure to permanently close a UST system that has been out of service for more than 12 months and failure to have required release detection. One hundred twenty-one owner/operators were cited for failure to register their UST system and 114 or 94% have complied and registered their tanks. Eighty-nine were cited for failure to permanently close a UST system that was out of service for more than 12 months and 70 or 79% of these have been properly closed. One hundred seventy-two release detection violations were cited and 150 or 87% have been corrected. Sixteen various violations of the Regulations were also cited.

The highest compliance rate was obtained from citations for failure to register. This is because it is relatively easy to register a system and there is no expense involved. The Bureau has a good compliance rate for release detection violations. Once owner/operators were informed of the regulatory requirements they complied. Also the petroleum industry has communicated to owner/operators that New Mexico is actively enforcing the Regulations and assessing penalties for noncompliance. Owner/operators are installing and maintaining release detection to avoid penalties. The lowest compliance rate is for violations regarding out of service USTs. These systems are usually abandoned and it is difficult to convince an uncooperative owner that he must remove a UST system at a site that has been out of service for a long period of time and is no longer producing an income for the owner. The five cases referred to the Office of General Counsel for further enforcement action fall in this category.

SUMMARY AND CONCLUSIONS

With traditional enforcement only 14% of the owner/operators cited complied with the USTR. This traditional enforcement program was too lengthy and cumbersome to be effective. The expedited enforcement (citation) program has been very successful for gaining compliance with USTR. By October 31, 1991, and eight (8) months into the program, 52% of the owner/operators cited complied with the Regulations for a total compliance rate of 86%. The Bureau will continue to utilize expedited enforcement. This citation and penalty program gives the owner/operator an incentive to comply. This program also allows the Bureau to do more enforcement actions, with less attorney and staff time. The inspectors are able to do more inspections and the legal staff is able to focus on the habitual offender and the recalcitrant owner/operator. Other Bureaus with the Environment Department are also examining this successful enforcement system. Other programs face the same problems the UST Bureau did with traditional enforcement; it's time consuming and costly. Expedited enforcement is an alternate enforcement tool that is less time consuming, less costly and more effective than traditional enforcement.

The purpose of an inspection is both education and enforcement. Inspections provide an opportunity for the inspector to inform owners/operators of their legal requirements, and further explain the release detection regulations. By implementing an inspection program to effectively educate owners/operators and enforce the Regulations, the UST Bureau has significantly increased compliance with the Regulations; and thereby reduced the threats to New Mexico's environment posed by petroleum products and hazardous substances released from underground storage tanks.

William P. Gulledge[1]

IMPACT OF STANDARDS AND CERTIFICATION ON ENVIRONMENTAL IMPAIRMENT LIABILITY INSURANCE PROGRAMS

REFERENCE: Gulledge, W. P., "Impact of Standards and Certification on Environmental Impairment Liability Insurance Programs," Leak Detection for Underground Storage Tanks, ASTM STP 1161, Philip b. Durgin and Thomas M. Young, Eds., American Society for Testing and Materials, Philadelphia, 1993.

ABSTRACT: Environmental impairment liability (EIL) insurance is available for petroleum storage tank and other environmental exposures. Recent standards and performance criteria for leak detection for underground storage tanks (USTs) and other technical standards for USTs have been both a benefit and an interference to risk-based underwriting of storage tank EIL insurance programs. Insurance underwriters and state financial responsibility program administrators are confronted with confusing information to manage these environmental risks.

Standards and certification are also key issues for site assessment programs. Recent activities from ASTM and the Institute for Environmental Auditing (IEA) have addressed the need to increase the professional stature of site assessments and environmental management. Reaction and acceptance of these efforts by the users have been mixed. Ultimately, these efforts will greatly impact insurance coverage for environmental risks.

KEYWORDS: environment, pollution, liability, insurance, underground storage tanks, certification, site assessments, Institute for Environmental Auditing

An insurance policy written specifically to cover pollution damages is a relatively new insurance product. During the early 1980s, over forty insurance companies offered EIL insurance either as an add-on to traditional property/casualty insurance or as stand alone coverage written on an "occurrence" basis. In 1985-86, the EIL insurance market turned sharply downward. Losses greatly exceeded premiums. The large losses can be attributed to a variety of reasons including:

1. Little understanding of the environmental risk being insured;
2. No method to translate environmental risk into a financial responsibility mechanism;
3. No specialized capability to manage environmental claims; and

---

[1]President, Institute for Environmental Auditing, Washington, D.C., and Consultant, Tillinghast, 4601 North Fairfax Drive, Suite 1100, Arlington, VA 22203

4. Poor construction of policy coverage.

The lack of environmental standards and certified environmental professionals or programs led, during the early and mid 1980s, many insurers to underestimate the complexity of underwriting environmental risks and to inadequately respond to pollution incidents. Profitable insurance coverage is normally written using a defined, well understood risk profile. Loss estimates can then be developed with a high degree of confidence in the results. Traditional insurers have considerable experience in estimating business liability risks, but until recently, designing environmental insurance programs or realistically estimating the environmental risks of site operations was not approached on a professional basis.

Pollution insurance developed in the late 1980s and today normally has used specialized expertise to construct at least certain portions of a comprehensive EIL insurance package. A better attempt by insurers to characterize the risk and ultimate losses is quite evident, but the lack of clearly understood technical standards has compromised their efforts. Insurance for petroleum storage tank risks is a good example.

FINANCIAL RESPONSIBILITY FOR PETROLEUM STORAGE TANKS

The U.S. Environmental Protection Agency (EPA) promulgated financial responsibility requirements for owners and operators of petroleum USTs in late 1988. EPA regulations require each owner or operator to demonstrate $500,000 or $1 million per incident of financial responsibility, depending upon whether petroleum products are sold to the public and the facility petroleum product volume. Insurance is one option for demonstrating financial responsibility.

A number of companies developed policies to meet the new requirements. Most companies also used part or all of the companion technical regulations as underwriting criteria to screen and/or price coverage for UST risks. Much of the underwriting focus has been placed on compliance with leak detection standards and especially the tank testing option to meet the leak detection standard.

However, most insurers have ignored the technical probability data for tank testing. Typical supporting information to underwrite UST exposures have included: tank tests, environmental site assessments, inventory reconciliation, and more simply just examining the age and construction of the UST. An example of simple underwriting criteria could be: "no bare steel USTs over 15 years of age unless the system is tested tight." If tank tests are used as the primary evidence that the UST is not leaking, the underwriter normally will accept any tank test. Consideration is rarely given to the testing methodology, environmental conditions of the location of the test, and expertise of the operator conducting the test.

Why are insurance companies not investigating or simply ignoring significant factors that create false positive and false negative results for tank tests? The perception of having a certified test operator is part of the answer. Lacking the specialized engineering expertise and physics knowledge to evaluate tank test applications and test operators, many insurers will put more credibility on an operator who is certified and a test methodology that has been verified by a third party. Two important facts are normally not considered by the insurer:

1. Certification criteria differ from state to state; and
2. Different test conditions are used to verify tank test methodology.

One may criticize this effort, but the fact that tank tests and other assessment data are being used to underwrite UST exposures represents a significant improvement over EIL underwriting efforts ten years ago. However, more comprehensive and meaningful standards and certification programs undertaken by government and professional associations could play a more complete role in underwriting. Leak detection alternatives offer an example.

It is well-known that many insurance programs reward better risks with lower premiums for a given amount of insurance protection. Life insurance is a good example. For a given insurance amount ($100,000), a non-smoker who regularly exercises has a cheaper premium than a smoker. Leak detection options for USTs can be based on an environmental risk profile. Assuming the same type of UST is located at two facilities, the facility that does a tank test using the cheapest methodology and practices inadequate inventory reconciliation would pay more for a $1 million per incident policy than the facility with an automated inventory control and vapor leak detection system.

A more focused, specific national standard that recognizes the true capabilities of each leak detection technology may enable more traditional insurers to enter the storage tank insurance market. States have also had difficulty with the vague criteria. Many states have gone beyond the EPA leak detection requirements and officially or informally recommend one option over another. Some states provide general information such as using vapor monitoring in non-saturated UST excavations and piping trenches or using groundwater monitoring in saturated conditions. Other states are recommending automatic tank gauging over external leak detection systems.

Some states and local governments are considering accelerating the compliance timetable required under the EPA technical rules. Under the federal requirements, owners are given between December 1989, and December 1993 (depending on the age of the UST system) to comply with the leak detection standards and until 1998 to meet the corrosion protection standard and spill and overfill prevention requirement. The ten-year period given to meet the overfill protection standard is the requirement most often challenged by state regulators. Several states are planning or have already reduced the ten-year compliance schedule.

State funds have assumed the financial responsibility of tank owners and operators in most states. State regulators are either acting as storage tank release insurers or cleanup administrators. The fact that several states have found national standards to be inadequate for either environmental protection or financial responsibility shows that at least some portions of the EPA UST regulations are not adequate for environmental risk management purposes.

Every state's financial resources are scarce. The existing national technical standards for UST regulation provide a broad conceptual reference for effective environmental risk management. State UST program administrators could make greater use of risk-based underwriting and financial responsibility tools (loans for upgrades, etc.) if either technical standards were more representative of the true performance of UST technology options or better technical data were distributed by EPA on a regular basis.

SITE ASSESSMENT AND CORRECTIVE ACTION STANDARDS FOR UNDERGROUND STORAGE TANK RELEASES

EPA decided to promulgate generic corrective action procedures. States are free to develop their own corrective action standards. The result has been a huge state-by-state discrepancy for cleanup levels that must be achieved for petroleum product releases, acceptable technologies that must be used to achieve the cleanup level, and administrative procedures that must be adhered to. The flexibility is certainly not all bad, and ASTM Subcommittee E-50.01 is in the process of developing a guide for corrective action procedures to simplify and accelerate the cleanup process.

The guide recognizes that site assessment and corrective action activities occur at many points in the remediation process. Each round of investigation leads to corrective actions which lead to additional investigation steps until corrective action goals have been achieved. A full site assessment is not needed. Interim corrective action measures can be used to address contamination as it is discovered.

The goal of corrective action is to reduce harmful exposures to public health and the environment. One objective of the ASTM standard guide is to accomplish this goal by reducing costs and time currently experienced in many remediation projects. Accomplishing this objective will provide insurers and state fund administrators more comfort that remediation and restoration costs can be controlled. The standard will benefit environmental risk management efforts.

SITE ASSESSMENT STANDARDS FOR COMMERCIAL PROPERTY TRANSACTIONS

ASTM Subcommittee E-50.02 has undertaken an effort to develop a standard guide for conducting Phase I environmental property assessments. The standard is intended to address only environmental exposures under CERCLA/SARA. The intent of this standard is for the user to qualify for the innocent landowner defense under CERCLA. Following the procedures outlined in the guide will enable the user to meet "all appropriate inquiry into previous ownership and uses of the property consistent with good commercial or customary practice" (42 USC 9601(35)(B)).

Phase I assessments are designed to identify the presence or likelihood of hazardous substances on a property. The assessment includes: a records review, site reconnaissance, interviews about the property, and preparation of a final report. Phase I assessments normally do not include any testing or sampling of materials.

Meeting "all appropriate inquiry" criteria is met by using a sliding scale of inquiry. The standard includes both a transaction screening process and a CERCLA-related Phase I environmental assessment. The screen is a less extensive inquiry that the Phase I process. If further investigation is warranted after preparing the transaction screen, a Phase I site assessment is conducted or the specific CERCLA exposure needing additional investigation is further evaluated. It is not necessary to prepare a Phase I document for properties that pass the transaction screen.

Standards are needed because of the confusion created in the commercial real estate industry, as well as in the consulting community, by vague statutory requirements for the defense and the lack of guidance from court decisions regarding liability of financial institutions for Superfund cleanup costs. Resulting environmental liability has expanded creating a compelling need for business to develop effective risk management strategies. The lack of standards for assessing environmental liabilities associated with commercial property transactions has hindered the development of EIL policies to cover this exposure.

New efforts are underway to develop EIL coverage from this standard and similar site assessment processes. Policies for environmental damages are currently available to lenders that would insure against mistakes or omissions made in the Phase I site assessment. These policies require that a complete Phase I site assessment be prepared prior to binding coverage for a specific property.

Other insurance companies are focusing on using both the transaction screen and the Phase I document as underwriting tools. Insurance for the environmental exposures of prior site activities as well as current and future site operations can be covered under these policies. The standards are important benchmarks for establishing a site's current environmental liabilities and for identifying potential future problems. A major insurance market for this coverage may develop within the next several years.

CERTIFICATION OF ENVIRONMENTAL PROFESSIONAL ACTIVITIES

While insurers will have a standard to use in underwriting insurance for commercial property transactions, certification of professionals using the standards is just beginning. The State of California has maintained a registry of environmental professionals (REP) for several years. REPs are not required to take an exam, but they are required to meet experience criteria for registration. Continuing education requirements have recently been added to the program.

The IEA is addressing a multi-tiered certification program. More expertise and experience is required to progress to each tier. Certification requirements will include membership status, training courses and experience criteria, examinations, and continuing education procedures. No "grandfathering" provisions will be available. Possible tiers include:

1. Lenders or their designates who prepare commercial property transaction screens;
2. Professionals conducting Phase I site assessments; and
3. Specialized environmental auditors or professionals who prepare environmental management program audits or compliance audits.

Certification of individuals who use these standards would provide insurers additional confidence in the work products being relied upon for underwriting EIL coverage. Both management initiatives for environmental protection- standards and certification- are still in their infancy, but characterizing environmental liabilities is moving in the direction of more established professions such as actuarial analysis and financial accounting. More comprehensive, cheaper, and usable EIL insurance policies will result from this effort.

AUDIT STANDARD FOR ENVIRONMENTAL MANAGEMENT PROGRAM EFFECTIVENESS

The IEA has also developed a final draft standard for evaluating an organization's environmental management system. The scope of the audit to be performed under this standard encompasses all policies, procedures and practices designed to identify and assess environmental issues. Adequacy and effectiveness are the keys to reviewing environmental management control.

This standard provides a guide to determining how effective an organization's overall effort is in controlling environmental risks. It is not a substitute for an environmental risk assessment or an insurance liability analysis. However, the standard does provide useful data to insurance underwriters.

Writers of general EIL insurance will realize the most benefit from this standard. The environmental management program audit furnishes data on how effective the organization controls both current and past environmental liabilities. Underwriters can translate this information into future loss estimates and EIL insurance premium pricing. The standard becomes an important contributor to both effective environmental management and financial responsibility demonstration.

CONCLUSION

The complexity of environmental regulation and environmental risk management have increased the need to develop substantive environmental assessment standards and certification programs. Several standards setting organizations are in the process of developing useful documents for environmental professionals. These standards will in turn be used by pollution liability insurance underwriters to determine insurability of environmental risks and to quantitatively estimate future environmental impairment liability.

Any standard or certification program is worthwhile only if it is accepted by those currently in the profession and it is perceived by non-professionals as providing confidence in the work products of those practitioners. Will the efforts of ASTM, IEA and other organizations be accepted? One is hopeful that they will be. Policy makers have recognized the need for these programs, and more importantly, they are beginning to recognize that standards and certification development is best left to professional associations, not the government.

Insurance for environmental exposures will become more widely available as standards and certification procedures are developed. Insurers are looking for stability in assessing risks and for confidence in those performing the assessments.

Site and Risk Evaluation

R. W. Hillger[1], J. W. Starr[2], M. P. MacArthur[3], and J. W. Maresca, Jr.[4]

CHARACTERISTICS OF NON-PETROLEUM UNDERGROUND STORAGE TANKS

REFERENCE: Hillger, R. W., Starr, J. W., MacArthur, M. P., and Maresca, Jr., J. W., "Characteristics of Non-Petroleum Underground Storage Tanks," Leak Detection for Underground Storage Tanks, ASTM STP 1161, Philip B. Durgin and Thomas M. Young, Eds., American Society for Testing and Materials, Philadelphia, 1993.

Abstract: It is generally acknowledged that a small fraction of the total underground storage tank population is used to store substances other than petroleum fuels. The detailed characteristics of these tanks, however, are not well known. Additional information is required if competent decisions are to be made regarding leak detection, tank upgrading, and tank management practices. In order to obtain more detailed information regarding these tanks, a survey was conducted to determine the primary features of tanks containing non-petroleum substances. Databases were generated that were based on information from 14 states covering a wide geographical area. The results of this survey suggest that, of the non-petroleum tanks, approximately 50%, either by number or tank volume, contain hazardous substances. Acetone, toluene, methanol, methyl ethyl ketone, and xylene were found to be the most commonly stored hazardous substances, comprising approximately 30% of hazardous materials stored in tanks. Tank age was found to average 18 years, with over 85% of the tanks being fabricated from steel. Roughly 60% of the tanks in the state databases had capacities between 1 000 and 10 000 gal (3 790 and 37 900 L), with the average tank size from all states being 7 205 gal (27 307 L).

Keywords: underground storage tank, hazardous chemicals, database, tank characteristics

INTRODUCTION

Federal underground storage tank (UST) regulations promulgated on 23 September 1988 (40 CFR Parts 280 and 281) establish a broad range of minimum requirements for

---

[1] Environmental scientist, U.S. Environmental Protection Agency, Releases Control Branch, Risk Reduction Engineering Laboratory, Edison, NJ 08837

[2] Manager, Edison Operations, and senior research engineer, Vista Research, Inc., 100 View St., Mountain View, CA 94041.

[3] Research engineer, Vista Research, Inc., 100 View Street, Mountain View, CA 94041.

[4] Vice president and staff scientist, Vista Research, Inc., 100 View Street, Mountain View, CA 94041.

the design, installation, operation and leak testing of UST systems, both those that contain petroleum and those that contain hazardous substances [1]. In the regulation the following definitions are used to distinguish petroleum UST systems from hazardous substance UST systems:

> *"Petroleum UST system* means an underground storage tank system that contains petroleum or a mixture of petroleum with *de minimis* quantities of other regulated substances. Such systems include those containing motor fuels, jet fuels, distillate fuel oils, residual fuel oils, lubricants, petroleum solvents, and used oils." (Page 37 196 of [1]).

> *"Hazardous substance UST system* means an underground storage tank system that contains a hazardous substance defined in section 101(14) of the Comprehensive Environmental Response, Compensation and Liability Act of 1980 (but not including any substance regulated as a hazardous waste under subtitle C) or any mixture of such substances and petroleum, and which is not a petroleum UST system." (Page 37 196 of [1]).

These regulations were designed to help the user community minimize the adverse environmental impact caused by leakage of product from USTs. The regulations specify the type of testing methods that are acceptable for leak detection and either a minimum performance standard for each method or a specific set of instructions on how to conduct a test. The leak detection standards for volumetric methods were developed specifically for petroleum UST systems through an extensive theoretical and experimental test program conducted at the U.S. Environmental Protection Agency's (EPA's) Underground Storage Tank Test Apparatus in Edison, New Jersey, on 8 000-gal (30 320-L) tanks containing unleaded regular gasoline [2-9]. These same standards can also be applied to hazardous substance UST systems under certain circumstances and with special regulatory approval. In general, the same test methods and standards can be applied to a hazardous substance UST system if the substance contained in the tank is considered to be no more toxic than petroleum if released into the environment and if methods exist to remediate the site in case such a release occurs [1].

Since a large portion of the federal standards for volumetric methods is based on data developed from only a single petroleum fuel, an issue that should now be addressed is the influence of the chemical and physical properties of hazardous substances other than petroleum fuels on the ability of leak detection methods to satisfy the mandated performance standards. It is generally acknowledged that hazardous substance UST systems contain a wide variety of liquid products having a broad distribution of chemical and physical properties. Additional information is required before the potential impact of these substances on the performance of volumetric leak detection methods can be assessed. This information includes the types of substances that are most commonly stored and the sizes of the tanks in which the substances are kept. Once this assessment has been completed, other decisions about tank upgrading and tank management practices can be made.

This paper describes the results of a survey conducted in 1990 of the characteristics of the non-petroleum UST systems registered in 14 states. Information derived from the survey was organized into 14 databases, one for each state. Eleven of the 14 databases include UST systems that contain both hazardous and non-hazardous substances (e.g., mineral spirits); none of the data presented in this paper contain information on petroleum UST systems. The primary objective of this study was to tabulate the important characteristics of non-petroleum USTs. With this information, an analysis of the applicability and performance of volumetric methods of leak detection was performed. This analysis is described in a separate paper [10].

An earlier survey of hazardous substance UST systems was conducted in 1987 [11]. The 1987 survey was limited in scope, and relied on data from two large states, California and New York, and a compilation provided by the Chemical Manufacturers Association (CMA). Since the CMA data are not broken down by geographical location or by tank size, it is difficult to make a direct comparison between the 1987 survey and the present one. Where appropriate, some of the pertinent results from the earlier survey are presented here.

The 14 states included in the 1990 survey are Delaware, Florida, Illinois, Indiana, Maine, Massachusetts, Minnesota, Missouri, Montana, New York, Ohio, Texas, Virginia, and Wisconsin. In selecting these states, efforts were made to obtain representative national coverage while simultaneously examining the more populous industrial states, which might be expected to have large numbers of hazardous substance UST systems. The data collected from this survey were based primarily on the responses to the national underground tank registration requirements, which were instituted in 1984 as part of the amendments to the Resource Conservation and Recovery Act (RCRA). For each state, information regarding stored substances was compiled; this included CERCLA name and/or CAS number, tank capacity, tank construction material, and tank age. The resulting data were then organized and sorted so that the basic characteristics of the sample population could be tabulated.

CHEMICAL DISTRIBUTION

Previous studies suggest that, of the total number of underground tanks installed in the United States, approximately 95% are considered to be petroleum USTs [1,11]. The remaining 5%, comprising non-petroleum USTs, are devoted to the storage of a vast array of hazardous and non-hazardous substances. The 1987 survey of data from California, New York, and the CMA indicated that hazardous substance USTs comprised slightly over half of the non-petroleum tank population [11]. The 3 766 hazardous substance USTs registered in California comprised 54% of the non-petroleum tanks and 2.25% of the total population of 166 973 tanks in the state. In New York 792 (1.07%) of the 73 819 registered tanks contained hazardous substances. Over 80% of the hazardous USTs tabulated in this previous survey contained organic substances.

Table 1 gives the percentage of hazardous and non-hazardous substances stored in USTs for each of the 14 states that participated in the present survey. The total number of non-petroleum USTs tabulated from the 14 state databases was 9 656; 5 528 of the tanks in these states contained hazardous substances. In the 11 states that registered tanks containing both types of substances, there was, as expected, a fairly large spread in the relative proportion of hazardous-substance tanks to non-hazardous-substance tanks, regardless of whether this was tabulated by the total number of tanks in each state or the total storage capacity of the tanks. The total number of non-petroleum USTs in these 11 states was 8 138, with 3 979 of the tanks containing hazardous substances. The average percentage of tanks containing hazardous substances in these 11 states (based on the total number of tanks in these states) was 48.9%. Based on the total volumetric capacity of the tanks in the 11 states, 46.9% of the non-petroleum USTs contain hazardous substances, which is similar to the estimate based simply on the number of tanks.

TABLE 1--Percentage of hazardous and non-hazardous substances stored in registered non-petroleum USTs.

State	% by Number of Tanks			% by Volume	
	Number	% Haz	% Non-Haz	% Haz	% Non-Haz
Delaware	18	78.9	21.1	97.0	3.0
Florida	404	28.2	71.8	30.4	69.6
Illinois	2 862	72.0	28.0	68.1	31.9
Indiana	506	100.0*	0.0*	100.0*	0.0*
Maine	667	28.8	71.2	22.6	77.4
Massachusetts	66	49.0	51.0	54.8	45.2
Minnesota	406	24.9	75.1	28.9	71.1
Missouri	734	33.0	67.0	47.4	52.6
Montana	76	100.0*	0.0*	100.0*	0.0*
New York	936	100.0*	0.0*	100.0*	0.0*
Ohio	795	35.6	64.4	43.6	56.4
Texas	692	57.4	42.6	57.1	42.9
Virginia	896	34.2	65.8	42.7	57.3
Wisconsin	598	44.8	55.2	49.5	50.5
Average		49.3**	50.7**	53.1***	46.9***

*   Denotes that only hazardous, non-petroleum chemicals were reported in the database for that state.
**   The average is computed using the total number of hazardous or non-hazardous tanks in the 11 states containing both types of substances divided by the total number of non-petroleum tanks in these states (8 138).
***   The average is computed using the volumetric capacity of hazardous or non-hazardous tanks in the 11 states containing both types of substances divided by the total number of non-petroleum tanks in these states (8 138).

The survey data were sorted to determine the most commonly stored hazardous substances. The results are shown in Table 2. These data were sorted by the percentage of the total number of hazardous tanks, the percentage of the total number of non-petroleum tanks, and the percentage of the volumetric capacity of the non-petroleum tanks. The sorting indicates that toluene, acetone, methanol, methyl-ethyl ketone and xylene are the most commonly stored hazardous substances and that each substance, with the exception of xylene, comprises between 5 and 10% of the total number of hazardous USTs. These five substances comprise approximately 30% of all of the hazardous substance USTs identified in the 14 state databases, a significant fraction of the total tank population. These same five hazardous substances were also identified as comprising the largest fraction of hazardous substances in the 1987 survey [11].

TABLE 2--Most commonly stored hazardous substances based on data from all 14 states.

Chemical	% Hazardous Substance Tanks[*] by Tank Number	% of Non-Petroleum Tanks[**] by Tank Number	% of Non-Petroleum Tanks[**] by Tank Volume
Toluene	9.8	5.6	9.2
Acetone	6.9	3.9	4.2
Methanol	6.7	3.8	3.3
Methyl-Ethyl Ketone	6.5	3.7	2.9
Xylene	---	---	2.5
TOTAL	29.9	17.0	22.1

* The data are reported as a fraction of the total number of all hazardous tanks.
** The data are reported as a fraction of the total number (or volume) of all non-petroleum tanks.

TANK CAPACITY

The range of tank sizes in the survey, as well as the number of tanks in different size ranges, are summarized in Table 3. The data were sorted into five different groups. Excluding Delaware, the results indicate that the average tank size in all surveyed states for which data were available ranged between 3 409 and 12 400 gal (12 920 and 46 996 L) and that the median tank size was 6 826 gal (25 870 L) (Illinois). The aggregate average tank size, which is based on an average of all of the tanks in the 14 states (i.e., 9 656 gal [36 596 L]), was found to be 7 205 gal (27 307 L). The largest size reported for an individual tank (found in Delaware) was 430 000 gal (1 629 700 L).

It is clear from these data that the majority of the tanks exhibit capacities of 20 000 gal (75 800 L) or less. In addition, over 70% of the tanks have capacities less than or equal to 10 000 gal (37 900 L), with the two largest groups comprising the range between 1 000 and 10 000 gal (3 790 and 37 900 L). With the exception of Delaware, the average tank volume for most states was generally found to be between 6 000 and 9 000 gal (22 740 and 34 110 L); this volume range brackets the average size of tanks storing these substances. The data for Delaware are comprised of only 18 tanks, four of which have capacities of 430 000 gal (1 629 700 L) each, resulting in an extremely biased average tank volume.

TABLE 3--Tank size distributions compiled from the 14 state databases and expressed as a per cent fraction of the number of tanks in each state.

State	Range of Tank Capacities (Percent of Tanks in Each State)					
	< 1 000 gal (<3 790 L)	1 000- <4 000 gal (3 790- <15 160 L)	4 000- <10 000 gal (15 160- 37 900 L)	10 000- <20 000 gal (37 900- 75 800 L)	>20 000 gal (>75 800 L)	Average Volume (gal (L))
Delaware	5.6	16.7	27.8	22.2	27.8	101 293 (383 900)
Florida	27.7	39.9	22.0	7.4	0.5	3 409 (12 920)
Illinois	7.1	29.2	33.6	19.8	5.9	6 826 (25 871)
Indiana*	4.3	16.0	26.5	19.8	29.6	11 525 (43 680)
Maine	6.2	26.2	36.9	24.6	6.2	8 226 (31 177)
Massachusetts	15.5	32.1	28.7	19.7	4.0	6 132 (23 240)
Minnesota	15.3	34.5	23.9	18.2	7.1	6 211 (23 540)
Missouri	10.1	28.7	31.9	21.0	8.3	9 144 (34 656)
Montana*	44.7	23.7	19.7	3.9	7.9	12 400 (46 996)
New York*	12.8	22.3	30.8	23.5	10.6	8 957 (33 947)
Ohio	6.6	33.8	37.9	18.1	3.6	5 546 (21 019)
Texas	11.3	28.0	28.9	19.7	6.8	6 952 (26 348)
Virginia	15.8	29.4	28.7	17.0	6.9	6 534 (24 764)
Wisconsin	8.4	30.4	37.6	19.7	3.8	6 350 (24 067)
AVERAGE	11.3	29.6	31.9	19.6	7.6	7 205 (27 307)

* Totals for New York, Indiana, and Montana are based on CERCLA chemicals only.

TANK CONSTRUCTION

Table 4 summarizes the material of construction for the non-petroleum USTs based on the data from all 14 states. The most common construction material is carbon steel. It is interesting that in some of the states a significant portion of the installed tanks are reported to be of unknown construction, with as many as 22.3% of the tanks in Florida fitting this category. Carbon steel was also found to be the most common construction material in the 1987 survey [11], comprising slightly over 90% of the population surveyed. Of particular note in this earlier survey was the large number of tanks in which corrosion protection was not employed or in which the type of protection employed was unknown; over 40% of the tanks exhibited this characteristic.

TABLE 4--Tank construction materials compiled from 14 state databases.*

State	Steel (%)	Fiberglass Reinforced Plastic (%)	Other (%)	Unknown (%)
Delaware	77.8	0.0	5.5	16.7
Florida	62.9	9.2	5.7	22.3
Illinois	89.4	4.2	2.6	3.8
Indiana*	---	---	---	---
Maine	72.7	15.2	10.6	1.5
Massachusetts	90.3	5.4	3.0	1.3
Minnesota*	---	---	---	---
Missouri	83.4	7.2	6.1	3.3
Montana	85.5	---	---	---
New York	83.9	12.3	3.8	0.0
Ohio	94.1	2.1	1.5	2.3
Texas*	---	---	---	---
Virginia	86.5	5.9	5.1	2.6
Wisconsin	79.3	9.4	7.2	4.2
AVERAGE	86.1	6.2	3.9	3.8

* Values reported are percentages of the total tank populations in each state. Materials were not reported for Indiana, Minnesota, and Texas. Only steel tanks were reported for Montana.

TANK AGE

Table 5 summarizes the tank age distribution as a percentage of the total tanks included in the 14 states. As can be seen in this table, the mean age of all tanks containing non-petroleum substances is 18.3 years, with nearly 40% of the tanks being over 20 years old.

TABLE 5--Tank age distributions compiled from 14 state databases and expressed as a percentage of the number of non-petroleum USTs in each state.

State	Range of Tank Age (Years)				
	0 to 4	5 to 9	10 to 14	15 to 19	>= 20
Delaware	22.2	5.6	5.6	0.0	66.7
Florida	3.5	31.4	18.7	11.0	35.4
Illinois	2.6	12.1	22.9	18.2	44.2
Indiana	6.9	15.2	20.6	16.8	40.5
Maine	9.1	14.1	13.9	16.2	46.7
Massachusetts	1.8	14.0	43.9	17.5	22.8
Minnesota	3.0	23.0	22.2	10.0	41.8
Missouri	3.3	20.9	23.0	16.0	36.8
Montana	2.7	4.0	4.0	2.7	86.7
New York	14.8	15.2	20.0	14.2	35.8
Ohio	1.5	15.8	20.4	21.5	40.8
Texas	6.2	21.3	26.1	17.9	28.4
Virginia	2.1	19.1	21.6	21.1	36.0
Wisconsin	4.3	13.8	24.9	25.6	31.5
AVERAGE*	4.9	16.2	21.6	17.6	39.7

* The average is computed using the total number of non-petroleum tanks in the 14 states.

DISCUSSION

A number of upper bound estimates of the total number (and percentage) of underground tanks containing hazardous substances in the United States was made from the state databases. While the number of petroleum USTs might be expected to be directly proportional to the population of each state, this would not necessarily be true of the hazardous substance USTs. However, one would generally expect the states with the largest populations and, therefore, the largest industrial bases, to have the greatest number of hazardous tanks. All of the estimates suggested that the maximum number of hazardous substance USTs would be between 10 000 and 20 000, or between 0.5 and 1% of all USTs. In estimating these percentages it was assumed that the total number of USTs was two million. It is not known exactly how many USTs there are in the United States, but the total number is large. As of 1988, the number of registered petroleum USTs was over 1.4 million [1]. The number of hazardous substance USTs in the nation was estimated by taking the data from both the 11-state and 14-state databases and extrapolating by the fractional population represented by those states or by the fractional number of states. Other estimates were made by partitioning the states by population into quartiles and by multiplying the number of states in each quartile by the state with the maximum number of hazardous substance USTs in each quartile or by the state with the maximum population in each quartile. Half the total population of the United States is found in nine states, and three of these, California, New York, and Texas, comprise approximately 25% of the total population. Between 2 and 4 states in the 14-state database sample were found in each quartile.

Based upon the above tabulations, the mean tank age was found to be roughly 18 years, with over 40% of the tanks being older than 20 years. The average tank capacity was found to be roughly 7 200 gal (27 288 L), with approximately 40% of tanks having capacities less than 4 000 (3 790 L) and approximately 40% having capacities over 20 000 gal (75 800 L). Over 86% are fabricated from steel. The average age of the non-petroleum USTs is approximately 6 years greater than those tanks containing petroleum, and the average capacity of the non-petroleum tanks is over 30% greater [12]. The use of inventory control and leak detection is more widespread for petroleum USTs than for non-petroleum USTs [13]. Based on the characteristics tabulated in this study and on previous experience with petroleum USTs [12,14,15], one might be tempted to conclude that a large portion of the non-petroleum UST population is at risk of failure (i.e., of developing leaks) due to corrosion. Whether this is true or not is very difficult to determine without conducting a detailed survey similar to the one for petroleum USTs conducted in 1985 [12]. A number of reports have addressed the causes of leaks [12,14,15]. Some of the pertinent findings from these reports are:

- Corrosion is the major failure mode for existing tanks [15].

- Small steel tanks are more likely to perforate than large steel tanks because they have thinner walls; tank size (i.e., wall thickness) is more important than age in predicting tank failure [14]; this finding, however, is not supported by data in EPA's national survey of underground tanks [12].

- Of the tanks studied in this survey, the average age of tanks with perforations due to corrosion was 23 years [14]. This was confirmed in another study [12], which found that steel tanks over the age of 20 years show substantial increases in releases.

Although information is limited, it appears that most of the owners of non-petroleum USTs are replacing or are planning to replace their single-walled tanks either with double-walled underground tanks or with aboveground tanks before the 1998 tank-upgrading deadline specified in the EPA regulations [13]. Volumetric test methods could be used to help verify the integrity of hazardous substance tanks until tank replacement is completed.

CONCLUSIONS

This survey tabulated the type of substances, by number and volume, contained in the non-petroleum USTs registered in 14 states. Approximately half of these are hazardous substance USTs, which, based on a simple calculation, represent less than 1% of all USTs in the United States (both those containing petroleum substances and those containing non-petroleum substances). The most striking feature of the surveyed non-petroleum UST population is the wide variety of substances that are stored. Among the hazardous substances contained in USTs, the most commonly stored are acetone, toluene, methanol, methyl ethyl ketone, and xylene; these comprise approximately 30% of the number of surveyed hazardous substance USTs.

In addition to assessments of the most commonly stored hazardous substances, assessments were made of the range of tank ages, materials of construction, and capacities. Based upon these analyses, the mean tank age was found to be 18 years, with over 86% of the tanks fabricated from steel. The average tank capacity was found to be roughly 7 200 gal (27 288 L), with over 60% of the tanks having capacities between 1 000 to 10 000 gal (3 790 to 37 900 L). In view of these findings, substantial upgrading of tank installations can be expected to occur over the next 6 years, as tank owners comply with the 1998 upgrading requirements in the federal regulations.

The data tabulated in this survey were used to determine whether volumetric leak detection methods offer a viable approach to testing the integrity of non-petroleum USTs [10]. The physical properties of a substance (particularly the coefficient of thermal expansion) determine the size of the volume changes that will occur in a *nonleaking* tank. The properties of the hazardous substances considered here are generally associated with volume changes that are not as large as those that occur in a tank containing unleaded gasoline, the substance on which the quantitative performance standards in the regulations are based. In addition, the average size and construction of the hazardous substance tanks closely approximate those from which supporting data for the regulations were developed. As a consequence, assuming that practical details of material compatibility and safety have been addressed, only minimal extrapolations of current knowledge should be needed in order to test the integrity of tanks containing non-petroleum hazardous substances.

REFERENCES

[1] U.S. Environmental Protection Agency, "Part 280 - Technical Standards and Corrective Action Requirements for Owners and Operators of Underground Storage Tanks," Federal Register, Vol. 53, 23 September 1988.

[2] U. S. Environmental Protection Agency, "Evaluation of Volumetric Leak Detection Methods for Underground Fuel Storage Tanks," Vol. I (EPA/600/2-88/068a) and Vol. II (EPA/600/2-88/068b), Risk Reduction Engineering Laboratory, U. S. Environmental Protection Agency, Edison, New Jersey, December, 1988.

[3] J. W. Maresca, Jr., "Volumetric Tank Testing: An Overview," Technology Transfer Report No. EPA/625/9-89/009, Center for Environmental Research Information, Office of Research and Development, U. S. Environmental Protection Agency, Edison, New Jersey, December, 1988.

[4] J. W. Maresca, Jr., J. W. Starr, R. D. Roach, and J. S. Farlow, "Evaluation of the Accuracy of Volumetric Leak Detection Methods for Underground Storage Tanks Containing Gasoline," *Proceedings of the 1989 Oil Spill Conference,* Oil Pollution Control, A Cooperative Effort of the U.S. Coast Guard, American Petroleum Institute and U.S. Environmental Protection Agency, San Antonio, Texas, March 1989.

[5] J. W. Maresca, Jr., J. W. Starr, R. D. Roach, J. S. Farlow, and R. W. Hillger, "Summary of the Results of EPA's Evaluation of Volumetric Leak Detection Methods," *Proceedings of the Fifteenth Annual Research Symposium,* Risk Reduction Engineering Laboratory, Office of Research and Development, U.S. Environmental Protection Agency, Cincinnati, Ohio, February 1990.

[6] J. W. Maresca, Jr., J. W. Starr, R. D. Roach, D. Naar, R. Smedfjeld, J. S. Farlow, and R. W. Hillger, "Evaluation of Volumetric Leak Detection Methods Used in Underground Storage Tanks," *J. of Hazardous Materials,* Vol. 26, 1991.

[7] J. W. Maresca, Jr., R. D. Roach, J. W. Starr, and J. S. Farlow, "U.S. EPA Evaluation of Volumetric UST Leak Detection Methods," *Proceedings of the Thirteenth Annual Research Symposium,* Hazardous Waste Engineering Research Laboratory, Office of Research and Development, U.S. Environmental Protection Agency, Cincinnati, Ohio, July 1987.

[8] R. D. Roach, J. W. Starr, C. P. Wilson, D. Naar, J. W. Maresca, Jr., and J. S. Farlow, "Discovery of a New Source of Error in Tightness Tests on an Overfilled Tank," *Proceedings of the Fourteenth Annual Research Symposium,* Risk Reduction Engineering Laboratory, Office of Research and Development, U.S. Environmental Protection Agency, Cincinnati, Ohio July, 1988.

[9] U. S. Environmental Protection Agency, "Volumetric Leak Detection in Large Underground Storage Tanks," Vol. I (EPA/600/2-91/044a) and Vol. II (EPA/600/2-91/044b), Risk Reduction Engineering Laboratory, U. S. Environmental Protection Agency, Edison, New Jersey, December, 1991.

[10] J. W. Starr, R. F. Wise, J. W. Maresca, Jr., R. W. Hillger, and A. N. Tafuri. Volumetric Leak Detection in Underground Storage Tanks Containing Chemicals. *Proceedings of the 84th Annual Meeting and Exhibition of the Air and Waste Management Association,* Vancouver, B.C., Canada, 15-17 June 1991.

[11] Lysyj, I., Hillger, R. W., Farlow, J.S., and Field, R., "A Preliminary Analysis of Underground Storage Tanks Used for CERCLA Chemical Storage," Proceedings of the Thirteenth Annual Research Symposium, Hazardous Waste Engineering Research Laboratory, Office of Research and Development, U.S. Environmental Protection Agency, Cincinnati, Ohio, July 1987.

[12] U. S. Environmental Protection Agency, "Underground Motor Fuel Storage Tanks: A National Survey," Vol. I, Technical Report (EPA 560/5-86-013) and Vol II, Appendices (EPA 560/5-86-013), Office of Pesticides and Toxic Substances, Washington, D. C., May 1986.

[13] R. F. Wise, J. W. Starr, J. W. Maresca, Jr., R. W. Hillger, and A. N. Tafuri. Underground Storage Tanks Containing Hazardous Chemicals. *Proceedings of the Seventeenth Annual Research Symposium*, EPA/600/9-91/002, Risk Reduction Engineering Laboratory, Office of Research and Development, U.S. Environmental Protection Agency, Cincinnati, Ohio, 3-5 April 1991.

[14] Pim, J. H., and Searing, J. M., "Tank Corrosion Study," Suffolk County Department of Health Services, Farmingville, New York, November 1988.

[15] U. S. Environmental Protection Agency, "Causes of Releases from UST Systems," (EPA UST 32A) Final Report, Office of Underground Storage Tanks, Washington, D. C., September 1987.

Appendix

Partial List of Chemicals Stored in UST

	CERCLA	NON-CERCLA
Ketones/Aldehydes	Acetone Methyl ethyl ketone Methyl iso-butyl ketone Cyclohexanone Formaldehyde	
Alcohols	Methanol n-Butanol iso-Butanol	Ethanol n-Propanol iso-Propanol Tridecyl alcohol 2-Ethyl hexanol 2-Methoxy ethanol 2-Ethoxy ethanol Methyl amyl alcohol Stearyl alcohol
Esters/Ethers/Glycols	Ethyl acetate n-Butyl acetate iso-Butyl acetate Dioctyl phthalate Ethyl ether	Ethylhexyl acetate n-Propyl acetate Trioctyl phthalate Cellosolve acetate Sodium octyl acetate Sodium phenyl acetate Methyl ether Propylene glycol-methyl ether Ethylene glycol Propylene glycol
Aromatic Hydrocarbons	Benzene Toluene Xylene	1-Propyl toluene
Chlorinated Hydrocarbons	Methyl chloride Methylene chloride 1,1,1, Trichloromethane Carbon tetrachloride Ethylene dichloride	Trichloromonofluoromethane
Monomers	Styrene Propylene oxide Vinyl acetate Methyl methacrylate Ethyl acrylate	Butyl acrylate
Miscellaneous Chemicals	Acetic acid Propionic acid Adipic acid Phenol Tetrahydrofuran Furfural Hydrazine Monomethyl amine Toluene di-iso cyanide Acetic anhydride Allyl chloride Phosgene Carbon disulfide	Methyl cellosolve Ethyl cellosolve Butyl cellosolve Naphthol Perchloroethylene-hydrofluoric acid Hexyl cellosolve Sodium silicate 1-Nitropropane

Inorganic Chemicals   Sodium hydroxide   Potassium fluoride
                          Potassium hydroxide   Calcium nitrate
                          Hydrochloric acid
                          Sodium hypochlorite
                          Sodium cyanide
                          Ammonium thiosulfate
                          Ferric chloride
                          Ferrous chloride
                          Chromic acid
                          Chlorine
                          Zinc
                          Chromium
                          Phosphorus
                          Ammonium hydroxide
                          Nitric acid
                          Sulfuric acid
                          Phosphoric acid
                          Ammonium sulfide
                          Ferrous sulfate
                          Hydrogen cyanide

Kevin W. Ferguson[1]

RISK ASSESSMENT TO AN INTEGRATED PLANNING MODEL FOR UST PROGRAMS

REFERENCE: Ferguson, K. W., "Risk Assessment to an Integrated Planning Model for UST Programs," <u>Leak Detection for Underground Storage Tanks</u>, <u>ASTM STP 1161</u>, Philip B. Durgin and Thomas M. Young, Eds., American Society for Testing and Materials, Philadelphia, 1993.

ABSTRACT: The U.S. Postal Service maintains the largest civilian fleet in the United States totalling approximately 180,000 vehicles. To support the fleets daily energy requirements, the Postal Service also operates one of the largest networks of underground storage tanks nearly 7,500 nationwide. A program to apply risk assessment to planning, budget development and other management actions was implemented during September, 1989. Working closely with a consultant, the postal service developed regulatory and environmental risk criteria and weighting factors for a ranking model. The primary objective was to identify relative risks for each underground tank at individual facilities. Relative risks at each facility were determined central to prioritizing scheduled improvements to the tank network. The survey was conducted on 302 underground tanks in the Northeast Region of the United States. An environmental and regulatory risk score was computed for each UST. By ranking the tanks according to their risk score, tanks were classified into management action categories including, but not limited to, underground tank testing, retrofit, repair, replacement and closure.

KEYWORDS: environmental risk, regulatory risk, leak detection, exposure potential, compliance, pilot testing, USPS Northeast Region

---

[1]Environmental Programs Engineer, Environmental Management Division, United States Postal Service, 475 L'Enfant Plaza, SW, Washington, DC 20260-6423

INTRODUCTION

The Northeast Region of the USPS consists of fourteen divisions which provide mail delivery service to the six New England states, New York, Puerto Rico, the Virgin Islands, and half of New Jersey. This is one of the most densely populated demographic areas of the world and is served by approximately 5,400 U.S. Post Offices. Because of climate, latitude, and the nature of mail delivery, there is a strong dependency for petroleum use as motor and heating fuels. The region is served by approximately 2,234 underground storage tanks which required research and investigation to determine status and integrity. Gasoline, diesel and waste oil tanks are the focus of this study.

METHODOLOGY

Advanced Planning

Working with a consultant, the Northeast Region's Environmental Program Specialist developed extensively detailed guidelines and protocol to complete the risk study. The advanced planning phase of the assessment included:

1) development of the advanced planning risk protocol
2) creation of a risk scoring system
3) development of criteria weighting factors,
4) pilot demonstration and testing in the field
5) completion of tank risk assessments in the Northeast Region

Field Preparation

The first planning task consisted of the formulation of the risk assessment questionnaire and protocol. The multi-purpose questionnaire enabled field inspectors to record and quantify environmental and regulatory factors associated with the risk posed by each tank. Pertinent factors included: site geology, hydrology, the age, construction design of the UST and its associated piping; UST location relative to property boundaries and water table; pattern of adjacent landuse (residential,commercial), surface and groundwater use; product stored in the UST; and federal,state and local UST requirements.

After formulating the survey questionnaire, a pilot field demonstration was conducted in the Boston area to evaluate the effectiveness of the risk assessment protocol. Early progress in the pilot study was closely monitored in order to refine and adjust the survey form and data collection process. Methodology employed during the field demonstration was judged effective and the demonstration was considered successful.

Scope of the investigation was then expanded to include the entire Northeast Region inventory of gasoline, diesel and waste oil tanks. Any other tank types identified at the designated facilities were also included in the study in order to complete the facility data base.

Literature Survey

The tank risk assessment consisted of two phases: a literature search and site inspection. The literature search was conducted for each postal facility site with data collection consisting of information gathered from water resource maps, geologic maps, plans, atlases, federal and state environmental reports and records. These data included soil type and permeability; depth to ground water, distance to wells, adjoining property use and distance to surface water or conduit. Federal state and local UST regulations were also researched, reviewed and evaluated to determine the regulatory risk of each UST.

Site Evaluation

Site investigations were conducted to confirm data collected during literature searches and telephone interviews. Each site visit involved interviews with knowledgeable USPS personnel who were queried as to the age of tanks, type and usage. Tank schematics were requested and, if available, reviewed for construction details, distance from each tank to nearest property line was noted, and the questionnaire was completed. Any other specific condition that might have had an adverse effect on the tank such as proximity to electric lines was also noted.

MANAGEMENT ACTION PRIORITIES

Based on information gathered during the field survey, each tank was assigned a score for environmental and regulatory risk. An assigned number quantified the risk producing factors relative to each tank. The higher the scoring number shown for the tank the greater the risk the tank presented. These scores were then used to rank the tanks in decreasing order of risk (Table 1).

Environmental risk values were weighted heavier than regulatory risk and therefore the environmental risk became the primary sorting factor. Regulatory risk was used as the secondary sorting parameter. If two tanks exhibited the same risk value, the regulatory risk values were compared

Table 1 - Manchester Division Risk Ranking List

Division	ST	MSC	Post Office	Unit	Facility ID#	Tank ID#	Tank(1) Type	Total Envrmnt Risk Score	Total Regultry Risk Score	Div Rank	ST(4) Rank	Reg Priority
Manchester	ME	Portland	Portland	VMF	226900001	0808	C	676.00	60.00	1	1	7	1
Manchester	NH	Manchester	Strafford	Main Office	328220001	0202	A	527.75	300.00	2	1	25	1
Manchester	NH	Manchester	Strafford	Main Office	328220001	0102	A	527.75	300.00	3	2	26	1
Manchester	ME	Portland	Auburn	Main Office	220240G02	0102	A	449.00	20.00	4	2	107	2
Manchester	ME	Portland	Auburn	Main Office	220240G02	0202	A	449.00	20.00	5	3	108	2
Manchester	NH	Manchester	Manchester	VMF	324800G01	0105	C	246.00	40.00	6	3	204	4
Manchester	NH	Manchester	Manchester	VMF	324800G01	0305	A	238.75	15.00	7	4	209	4
Manchester	NH	Manchester	Manchester	VMF	324800G01	0205	A	238.75	15.00	8	5	210	4
Manchester	ME	Portland	Portland	VMF	226900001	0108	A	118.25	5.00	9	4	239	4
Manchester	ME	Portland	Portland	VMF	226900001	0708	A	118.25	5.00	10	5	240	4
Manchester	ME	Portland	Portland	VMF	226900001	0608	B	117.50	5.00	11	6	241	4
Manchester	ME	Portland	Portland	VMF	226900001	0408	A	TP	TP	12	7	292	4
Manchester	ME	Portland	Portland	VMF	226900001	0508	A	TP	TP	13	8	293	4

(1) A = Gasoline
    B = Diesel
    C = Waste Oil
    D = Motor Oil
    E = Heating Oil

(2) High Environmental Risk (>390) and High Regulatory Risk (<55)
    High Environmental Risk (>390) and Low Regulatory Risk (<=55)
    Low Environmental Risk (<=390) and High Regulatory Risk (>55)
    Low Environmental Risk (<=390) and Low Regulatory Risk (<=55)

(3) LE = Lease Expired
    TC = Tank Closed
    TP = Tank Pulled
    WD = Wrong Tank Type Designation
    NE = Does Not Exist

(4) This Division covers parts of both NH and ME.
    The rankings shown are related to the state in which the facility is located.

for the tie breaker. After ranking the tanks by risk, they were subdivided into four action priority categories. Priority categories are as follows:

PRIORITY 1: High Environmental Risk (>390) and High Regulatory Risk (>55)
PRIORITY 2: High Environmental Risk (>390) and Low Regulatory Risk (<=55)
PRIORITY 3: Low Environmental Risk (<=390) and High Regulatory Risk (>55)
PRIORITY 4: Low Environmental Risk (<=390) and Low Regulatory Risk (<=55)

Making the decision as to what made a high environmental risk was more difficult than establishing the regulatory risk threshold for categorization. Tanks that were discovered leaking during the survey were given the greatest environmental risk score and highest priority. A decision was made as to what combination of major factors associated with an UST would constitute high risk. The major factors of concern for USTs were: tank and piping, age, construction; corrosion protection; proximity to ground water; distance to the nearest drinking water resource; proximity to boundary and the distance to any surface water or drain conduit. Each of these 4 factors was evaluated to determine what condition would correspond to a high environmental risk. A composite of the high risk condition relative to each factor was made and entered onto the risk assessment form. The high risk scenario corresponded to a single walled, coated steel tank and piping without cathodic protection that was greater than 14 years old; less than 40 feet from the groundwater table; within 1300 feet of a water well and was within 100 feet of surface water or conduit. The environmental risk score associated with this tank was 390. Any tank exhibiting a score equal to or greater than 390 was rated a high risk; any score less than this value was designated as having a low environmental risk.

Regulatory risk was established by calculating the maximum allowable score which corresponds to all regulatory risk categories being in compliance for one year. This calculated value for maximum acceptable risk was 55. Any regulatory risk score greater than 55 indicated that the tank is presently out of compliance and deemed to be a high regulatory risk. Conversely, any tank which exhibited a value of less than 55 was deemed to be a low priority risk.

Data Manipulation and Analyses

To obtain a clear and accurate overview of UST conditions for the entire region, all data was entered into a Lotus spreadsheet data base. Using the data sorting capabilities of Lotus, the consultant was instructed to compile and display the UST data into three categories by priority. These categories were 1) Region 2) State and 3) Postal division.

Regional Summary

All data collected during this study were sorted, risk ranked, prioritized, output and displayed in tabular form and on a map for the entire region (Figure 1). All tanks in priority 1 were scheduled for replacement during fiscal year 1990. Tanks identified as priority 2 were scheduled for replacement immediately following those identified as priority 1 during 1990, if physically possible. Tank projects in both categories 1 and 2 were fully funded throughout the Northeast Region. An underground tank testing program was initiated simultaneously within both categories and later expanded to the entire inventory. Construction contracts were developed and issued one week after completion of this study. Tank gauging and precision tank testing although redundant; served to enhance data reliability.

State Summary

A state summary of underground storage tank risks was also necessary because the Northeast Region is affected by UST regulations of eight states as well as federal regulations that apply to the territories of the United States. Table 2 is a synopsis of the applicable state and federal regulations that apply to existing USTs. Four postal divisions have post offices that span the boundaries of two states, in each case the state in which the post office is located will determine which regulations will apply. Tracking for compliance on a state by state basis facilitates improved reporting and permitting actions. It also fosters better program administration.

Division Summary

Providing a division summary of underground tank risks to local coordinators enabled them to develop annual and five-year plans to complete the program. Additionally, projects could be tracked by coordinators with greater ease and a better sense of direction. All tanks ranked in priorities 3 and 4 have been scheduled for annual testing and replacement to meet the double walled specification of the U.S. Postal Service prior to 1995. A typical map displays tank risks for the Manchester division (Figure 2).

FERGUSON/RISK ASSESSMENT TO AN INTEGRATED PLANNING MODEL   195

Table 2 - Regulations for Existing Petroleum Underground Storage Tanks

State	Applicability	Precision Tightness Test Requirements (Years After Installation)	Spill Containment Retrofit Requirements	Overfill Protection Retrofit Requirements	Cathodic Protection Retrofit Requirements	Closure Procedure	Mandatory Closure Dates for Non-Conforming Tanks	Maximum Out-of-Service Time Before Closure is Required
Connecticut	Motor fuel and waste oil	12, 13, 14, 15, 16, 17, 18, 19, closure year 20	12/22/98 (USEPA)	12/22/98 (USEPA)	12/22/98 (USEPA)	Removal or abandonment in place	Age 20	3 Months
Maine	All underground oil storage facilities	20, then every 5 years until closure	12/22/98 (USEPA)	12/22/98 (USEPA)	12/22/98 (USEPA)	Removal	1.	14 Months
Massachusetts	Motor fuel	10, 12, 15, 17, 19, then every year	May 30, 1990	May 30, 1993	12/22/98 (USEPA)	Removal	None	6 Months
New Hampshire	Any facility where the capacity of any one tank is more than 1,100 gallons	Now, 5, 10, 15, then every 5 years until closure	12/22/98 (USEPA)	September 1987	12/22/98 (USEPA)	Removal or abandonment in place	November 9, 1989: 24+ years old, all others by age 25	12 Months (USEPA)
New Jersey	Motor fuel and waste oil	None	September 1991 (pending)	September 1991 (pending)	September 1991 (pending)	Removal	None	12 Months (USEPA)
New York	Motor fuel	10 then every 5 years	12/22/98 (USEPA)	12/22/98 (USEPA)	12/22/98 (USEPA)	Removal or abandonment in place	None	12 Months (USEPA)
Rhode Island	Motor fuel and waste oil	5, 8, 11, 13, then every year	1987	12/22/98 (USEPA)	12/22/98 (USEPA)	Removal or abandonment in place	None	12 Months
Vermont	Motor fuel over 1,100 gallons and waste oil (any size)	10, 15, 19, 22, 25, then every year or 10, 13, 15, 17, 19, 21, 23, 25, then every year (sensitive area)	12/22/98 (USEPA)	12/22/98 (USEPA)	12/22/98 (USEPA)	Removal (pipes can be abandoned in place)	None	12 Months

1. Motor Fuel October 1, 1991: Age 25, Age 15 (sensitive area); October 1, 1994: Age 20, Age 15 (sensitive area); October 1, 1997: all non-conforming tanks. Waste Oil Age 20 + 2 months or warrantee date.

CONCLUSION

The following conclusions have been drawn from a review of the data collected during this study:

* 21% of the tanks surveyed fall into the PRIORITY 1 category
* 31% of the tanks surveyed fall into the PRIORITY 2 category
* 3% of the tanks surveyed fall into the PRIORITY 3 category
* 45% of the tanks surveyed fall into the PRIORITY 4 category
* No evidence of cathodic protection on any tank was found
* Only twenty tanks were found to have spill and overfill protection and spill containment devices in place
* Eight tanks surveyed during this study did not contain the liquids which they were reported to contain. The tanks held either heating oil or motor oil.
* Ten tanks on survey list were not located at the facilities where they were reported.
* Six tanks on survey list had already been removed or abandoned in place (approved closure method).
* The leases on four facilities held by the USPS had expired and had not been renewed.
* Six tanks in this survey were confirmed to be leaking.
* Twenty-one tanks were found to be in violation of the applicable state UST regulations which require their removal or abandonment (approved closure in place).

RECOMMENDATIONS

It is the author's opinion that the methods of evaluating environmental risk described in this study were useful for priority planning management actions associated with underground storage tank testing, repair, retro-fits, replacements, conversions and closures. This method of risk assessment coupled with ongoing manual tank gauging, inventory reconciliation, and precision tank tightness testing demonstrated a level of increased certainty relative to environmental and regulatory risk. By identifying high risk tank sites, more accurate measurement of program costs, contracting and equipment needs were ascertained; each of which enabled the Postal Service to launch a model Federal agency program. A determination of needs and economic decision analysis was also integrated with risk assessment results on an individual basis case by case. Since compilation of the risk assessment there have been 1534 UST projects initiated in the Northeast Region. State of the art automatic tank gauging systems, double-walled tanks, double wall piping and electronic leak detection equipment have been widely adopted throughout the nation by the U.S. Postal Service.

RESULTS

The Results section has been divided into two main sections. The first section deals with the distribution of Total Volatile Hydrocarbons (TVHC) and their relationship to various analytical values and physical site properties, and the second section deals with the relationship among various physical properties of the sites.

Most of the data has been presented on two different types of plots; correlation plots and histograms. In these plots whenever average concentrations were used they were derived from detected values only. This is because the true value cannot be estimated below the MDL.

On the correlation plots, TVHC concentrations have been plotted on log scales. On the histograms, TVHC is plotted on the abscissa and the individual ranges cover one log unit. For example, on the histogram in Figure 5, approximately 0.4% of the points had a TVHC concentration in the range greater than or equal to 0.01 μg/l to less than 0.1 μg/l and just over 16% of the points had a TVHC concentration in the range greater than or equal to 10,000 μg/l to less than 100,000 μg/l.

TVHC DISTRIBUTION

Detected TVHC concentrations ranged over 9 orders of magnitude, while individual detection limits varied over nearly 4 orders of magnitude. As was mentioned in the background section, the range of detection limits for TVHC values is predominantly a function of the size of sample injected into the GC, and to a lesser extent the condition of the GC or the analytical columns.

Shown in Figure 5 is a histogram depicting the distribution of TVHC values for all points in the Iowa database. Percentage of points for a given concentration range have been used in this plot, as in most of the histograms, to aid in normalizing the data.

As shown in the Figure, the histogram of detected TVHC values is asymmetric. This is because at the higher TVHC concentrations, the soil gas is approaching saturation. TVHC saturation is the concentration of hydrocarbon vapors in air immediately adjacent to the liquid hydrocarbons. The air inside the underground storage tank is near saturation. After this level is reached the vapor concentration of the hydrocarbons can no longer increase. TVHC saturation in the soil gas indicates the presence of a separate phase of hydrocarbons adsorbed to the soil particles is likely. Saturation concentrations vary according to the vapor pressure of the analyte. The plot of data from the unsaturated zone reflects the more random nature of that data while the TVHC values of approximately 10,000 μg/l and greater represent the proximity to an upper limit.

TVHC concentration was at or near saturation at 27% of the points and less than the MDL at 38% of the points. This leaves 35% of the

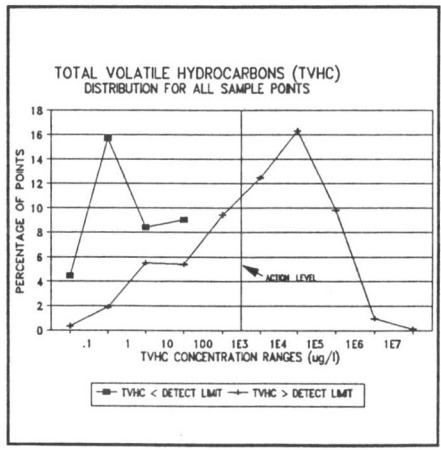

FIG. 5-TVHC distribution for 4195 sampling points throughout Iowa. Note asymmetry of plot as TVHC concentration approaches saturation. (1000 μg/L action level for state assisted cleanup has been added to Figure 5-8).

sampling locations with individual hydrocarbon concentrations between these extremes.

Figure 6 is a histogram of the probability distribution of the highest TVHC point for each site. As only the highest concentration at the site is being considered, the percentage of points with a concentration greater than 10,000 µg/l has increased significantly. Also, the percentage of points below the detection limit has decreased to 9.8% of the total and reflects the percentage of sites where no TVHCs were detected.

This histogram has the same asymmetric shape as seen when dealing with individual point data. This is useful for determining an action level, or fail threshold for a site. The predetermined action level for sites in Iowa was 1000 µg/l.[1] This figure is very conservative, and resulted in the failure of 72% of the sites in the initial screening process. A fail threshold of 10,000 µg/l would have resulted in the failure of 61% of the sites and probably would have been just as acceptable. The reason for this higher threshold is all samples were taken at potential source locations. In other words, if contamination was present in a level high enough to be of concern, it would be close to saturation at the source.

To determine the areal distribution of contamination it is necessary to use only the data from sites where leak detection was performed as these are the only sites where sampling locations were placed in the vicinity of the entire UST system. The histogram in Figure 7 shows the distribution of all sampling locations for leak detection sites. Approximately 35% of the sampling locations indicated hydrocarbon levels near saturation. On Figure 8; the histogram of the probability distribution for the highest point for each leak detection site; 74% of the sites are near saturation in at least one of their sampling locations.
Converting to absolute values, this means that 715 sampling locations are near saturation at 172 leak detection sites, or approximately 4 locations per site. With an average of 8.6 locations per site, this implies that 46% of the subsurface

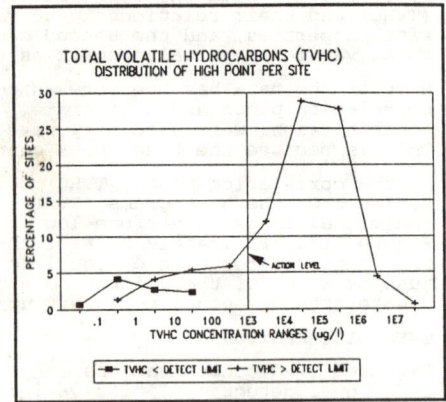

FIG. 6-Distribution of highest TVHC level detected at each of the 635 sites evaluated. Asymmetry more pronounced for site data as only highest values are used, therefore a greater percentage are near saturation.

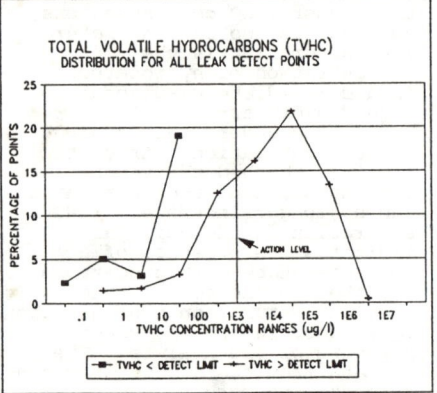

FIG. 7-TVHC distribution for 2012 sampling points at sites where UST systems were tested for leakage. Average TVHC concentration here is higher than the average for all points. (Figure 5).

near the underground fuel systems has soils containing hydrocarbons in a separate phase.

TVHC CONCENTRATION AS A FUNCTION OF VARIOUS PHYSICAL PROPERTIES

Three site characteristics were found to affect the TVHC distribution. Figures 9 through 15 illustrate the relationship between TVHC concentration and depth of sample, depth to groundwater, resistance to air flow in the soil, regional distribution and ground cover.

A common observation often reported in soil gas field data is increasing TVHC concentration with depth.[3] In many cases the source of TVHC's in the soil gas is TVHC contamination in the groundwater, and naturally, the closer the sampling location is to the source, the higher the concentration. The correlation plot of sample depth vs TVHC concentration for 2822 points in Figure 9 appears to agrees with this trend. The mediocre correlation coefficient of 0.44 is attributed to the samples being collected in the original source areas. What is probably being seen is a competition of the effect of vapors migrating up from the groundwater and vapors from more recent contamination of the soil in the tank pits and pipeline trenches. Also excess product will always be carried deeper into the soil until it reaches the water table. Therefore this trend might be observed whether the contaminants had reached the water table or not.

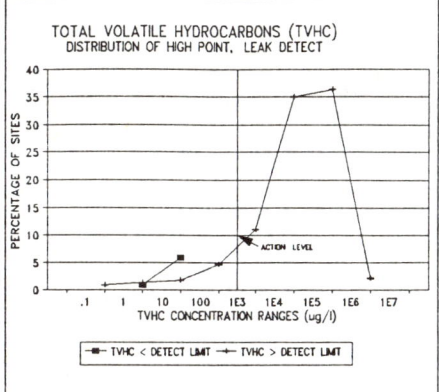

FIG. 8-Distribution of highest TVHC level detected at each of the 234 sites where UST systems were tested for leakage.

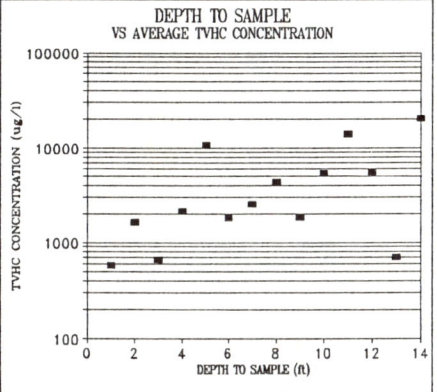

FIG. 9-Correlation between sample depth and TVHC concentration showing increase in soil gas contamination with depth. Correlation coefficient is 0.44 (2822 points). (1ft = 0.305m)

As would be expected from the previous discussion, there is a good correlation between the vertical distance from sample to groundwater and the average TVHC concentration. This relationship is illustrated in Figure 10, with the TVHC concentration decreasing as the distance to groundwater increases. At first the quality of the correlation is not apparent, however when the data for the 1107 points is weighted and a regression analysis is preformed, it is discovered that the correlation coefficient is 0.80. This is nearly twice as good as the TVHC vs Depth correlation. This is strong evidence that the water table was active as a barrier to the downward migration of hydrocarbons at a large number of contaminated sites in this study.

Relative permeability can be looked upon as the ease of fluid or air flow through a medium. The vacuum needed to draw a sample into the probe is an indication of resistance to air flow in the soil and is therefore a rough measure of the relative soil permeability: the higher the vacuum, the higher the resistance and the lower the permeability. Therefore, if the hydrocarbon contamination was a recent occurrence, an inverse relationship could be expected between the level of TVHC contamination and the resistance to flow in the soil. However, if the hydrocarbon contamination has been present in the soil for many years, the distribution would have been allowed to reach steady state, and a correlation plot would show little change in TVHC concentration.

As can be seen in Figure 11, the average TVHC concentration varies only slightly over the full range of flow resistance and thereby the full range of relative soil permeabilities. The conclusion would then be that contamination at the majority of sites is not a recent occurrence, but has had time to approach steady state. This conclusion is also supported by the histograms in Figures 5 through 8 which show that an upper limit is being reached.

For the purpose of determining regional variances in TVHC contamination, Iowa was divided into five areas based on zip code (Figure 12). No consideration was given to surface geology, as all the samples were taken from disturbed or non-native soil in tank pits or pipeline trenches. No significant regional variations in TVHC contamination were detected (Figure 13). The significance here is that this study is more universally applicable to places outside Iowa. Given that there is sufficient variation in the surface geology of Iowa, this is not reflected in the hydrocarbon distribution due primarily to sampling in the tank pits and pipeline trenches. In fact, the shape of the histogram for each area in Figure 13 correlates extremely well with the histogram of all TVHC points in Figure 5.

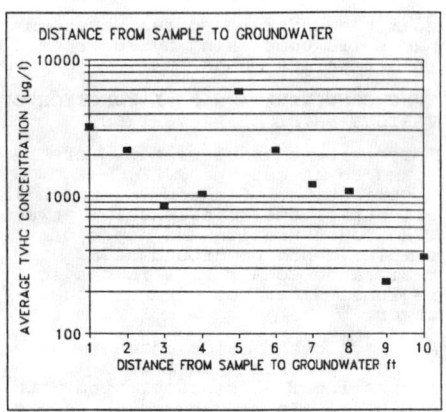

FIG. 10-Correlation between distance from sample to groundwater and TVHC concentration showing increase in contamination as groundwater is approached. Correlation coefficient of 0.80 supports the expected results. (1107 points) (1 ft = 0.305m).

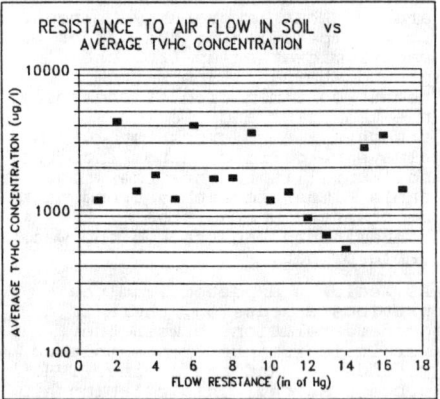

FIG. 11-Correlation between resistance to air flow in soil (relative permeability) and TVHC concentration. Almost no predictablity, correlation coefficient = 0.14. Hydrocarbons have had time to migrate into areas of low relative permeability indicating long term contamination. (1 in of Hg. = 0.3386mb).

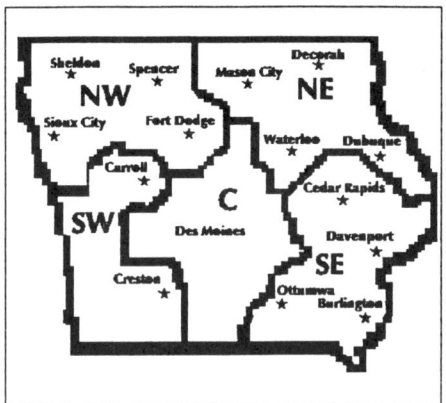

FIG. 12-Map of Iowa showing the areas used for regional TVHC distribution. No concern for surface geology as samples were taken from tank pits and pipeline trenches.(See Figure 13)

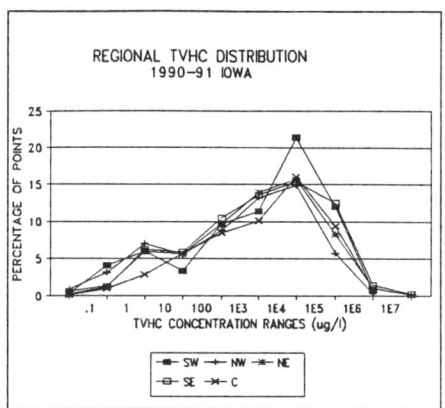

FIG. 13-Comparison of regional TVHC distribution for Iowa. No regional differences were expected as all samples were taken from non-native or disturbed soils.

Another physical property of the site which has an effect on TVHC concentration is ground cover. The relationship between ground cover and TVHC concentration is illustrated by the histogram in Figure 14 as well as the bar graph in Figure 15. Not surprisingly, a lower concentration of volatile compounds in the soil gas is found under unpaved areas for several reasons. First there would be no barrier to

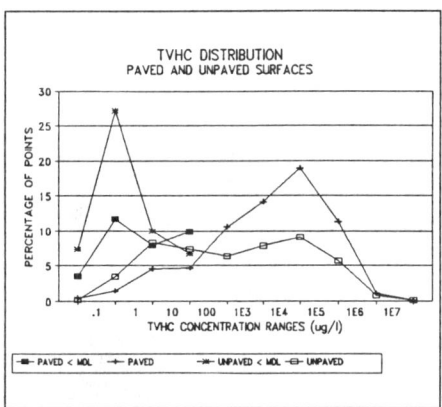

Figure 14-Distribution of TVHC concentration under paved and unpaved surfaces. Higher percentage of points <MDL and a lower percentage of points in the higher ranges due to weathering and biological activity where not protected by pavement.

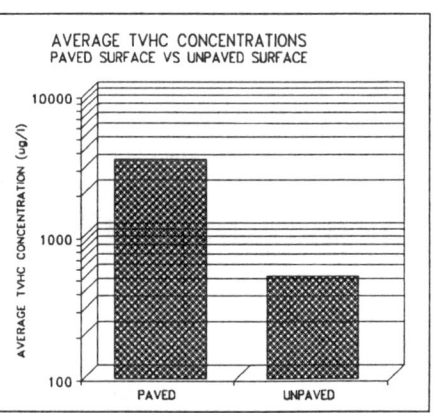

FIG. 15-Comparisons of average TVHC concentration under paved and unpaved surfaces.

prevent any of the volatiles from venting to the atmosphere. Secondly, and more importantly, no barrier to prevent oxygen from entering the soil. This oxygenated zone would allow for the growth of bacterial colonies which break down the hydrocarbons. Additionally, an unpaved area could act as a recharge zone for groundwater, periodically flushing the vadose zone and decreasing the amount of volatile compounds in the soil gas.

Note that the pattern for the histogram of TVHC concentrations under pavement, which accounts for 74% of the sampling locations, is an excellent match to the histogram of all TVHC points in Figure 5. Likewise, as demonstrated in Figures 13 and 16, the histograms of TVHC concentration vs resistance to air flow in soil and regional distribution all match well to Figure 5. The significance of this is TVHC contamination in the study area is fairly constant statewide and has been in place long enough to approach steady state. It is critical to remember that this consistency is probably due to the samples being taken from the most probable original source area.

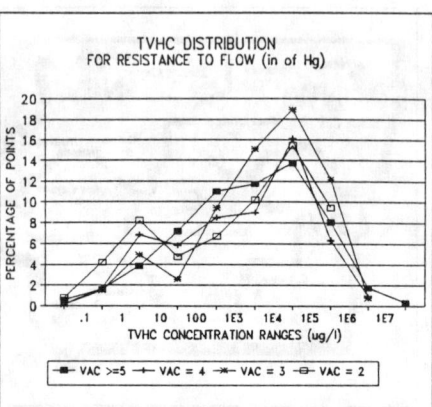

FIG. 16-TVHC distribution for various levels of resistance to air flow in the soil. Resistance measured as vacuum in inches of mercury. (See Figure 11) (1 in of Hg = 0.3386 mb)

Another factor to consider is whether TVHC contamination at a site is a reliable indicator of the integrity of the underground storage tank system. Or in other words what is the correlation between the test method used throughout the study and the concentration of hydrocarbons at the respective sites.

The test which was used to determine whether the UST system was tight utilizes a volatile and easily detectable chemical which is not normally found at UST sites. This chemical, called a tracer, was added to the UST system at a known concentration. After a period of time to allow the chemical to diffuse through the entire system and escape through any leaks, soil gas samples were taken from the tank pit and pipeline trenches, and analyzed for the tracer. The presence of the tracer in the soil gas at a concentration of 10^{-5} times the original concentration in the system, constituted a failure of the tank or pipeline.

Figure 17 shows the relationship between contaminated sites which passed a tightness test and those which failed a tightness test. Approximately 58% of the failed sites have high TVHC concentrations in excess of

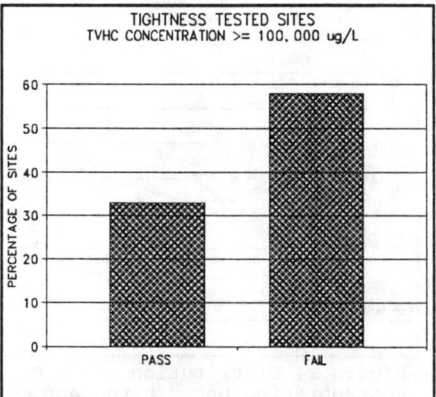

FIG. 17-Comparison of sites where a test for leakage was performed and the maximum TVHC value was greater than or equal to 100,000 µg/L. The odds are two to one of finding a leak at a site with very high levels of contamination.

100,000 μg/L, while 33% of the passed sites have high TVHC concentrations in that range. This would mean that when very high TVHC concentrations are detected, the odds of encountering a failure are two to one.

Low hydrocarbon readings at a site proved to be even more reliable as a prediction tool for determining if a system was tight. Of the sites which were tested, 16% had hydrocarbon concentrations less than the action level of 1000 μg/L. Of these sites only 8% failed the tightness test. Therefore, at sites where there are low levels of contamination, there is a 92% likelihood that the system will be tight (Figure 18).

RELATIONSHIPS AMONG SITE PROPERTIES

In order to insure that the results that were seen are due only to factors being considered, it is important to investigate possible relationships among several physical properties of the site which may have given false correlations in previous plots.

Figure 19 is a plot of the average resistance to air flow versus depth. As was previously mentioned, this resistance to flow correlates with the soil permeability and is measured in the field as vacuum required to take a sample, at a given flow, so produced by the pump. There is a bimodal correlation between these two properties. A slightly negative slope from 1 to 7 feet (0.3m to 2.13m) deep, and a strongly positive slope from 7 to 13 feet (2.13m to 3.96) deep. Although the upper 7 feet (2.13m) of soil is the zone most subjected to weathering, this could not account for the higher permeability in this zone as most of the points are under pavement.

The backfill in the tank pits is most probably composed of disturbed native soil or non-native soil. This disturbed zone will no longer reflect the natural changes in permeability with depth, but will probably be more homogeneous in its properties. This being the case, little change in permeability would be expected unless some other factor came into play. The

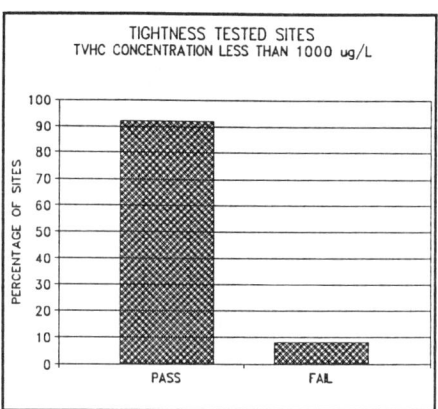

FIG. 18-Comparison of sites where a test for leakage was performed and all TVHC concentrations were < 1000 μg/L. The probability of a tight UST system is 92% where only low levels of contamination are found.

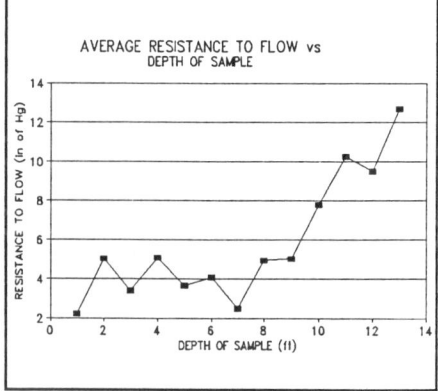

FIG. 19-Correlation of resistance to flow in soil (relative permeability) and sample depth. Note bimodal nature of correlatation, with break at depth of 7 feet (2.13m). Break due to proximity to groundwater in the capillary fringe. (See Figure 20)
(1 ft = 0.305m)
(1 in of Hg = 0.3386mb)

increasing resistance to flow from
7 to 13 feet (2.13m to 3.96m) is
probably due to the proximity to
groundwater which would have the
effect of increasing the resistance
to flow. The capillary fringe in
Iowa can be expected to be quite
thick as the native soil is
composed primarily of fine grained
glacial deposits (clays and silts).

This is supported by the available
groundwater data. Where
encountered, the average depth to
groundwater in all areas of Iowa is
approximately 7 feet (2.13m)
(Figure 20). However, groundwater
was encountered at only 164 of the
635 sites, and the average depth
reached without hitting water is
just less than 7 feet (2.13m).
Given that water is at 7 feet
(2.13m) or deeper over the entire
data base, the decreasing
permeability with depth beginning
at 7 feet (2.13m) is most likely
due to the proximity of
groundwater in the capillary
fringe.

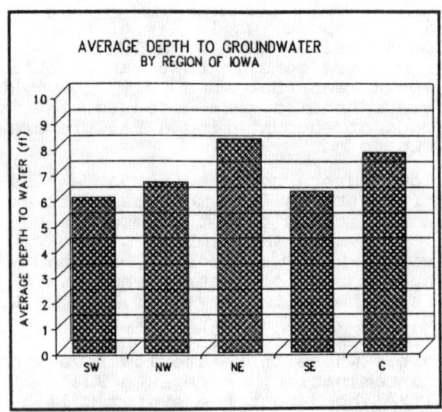

FIG. 20-Average depth to
encountered groundwater for the
5 previously defined areas of
Iowa. Average depth to
encountered groundwater
statewide was 6.95 feet (2.12m)
(1 ft = 0.305m)

CONCLUSIONS

After a review of the data it can be concluded that the use of soil
vapor analysis is a very useful tool in the assessment of UST sites.

TVHC concentration is approaching saturation at 61% of the sites in
Iowa. Of the sites which were tested for tightness and passed, 70% are
approaching soil gas saturation. Therefore the saturation of soil gas
with TVHC contamination is typical. However, when 100,000 μg/L TVHC or
higher is encountered, the odds of a failure in the system is two to
one.

Low hydrocarbon levels are a fairly reliable predictor that the
underground fuel system is tight. Only 8% of the sites with low
hydrocarbons failed a tightness test.

The distribution of hydrocarbon contamination varies with several
physical properties of UST sites. TVHC levels in the soil increased
with depth and decreased as the distance to groundwater increased.
Care should be taken when comparing samples taken under pavement to
samples from unpaved areas. The effect of biological activity as well
as flushing and aeration all have the effect of lowering the
hydrocarbon contamination where the soil is not protected by pavement.
As TVHC concentration did not vary with changes in the resistance to
flow in the soil, it can be concluded that at most of the sites, the
contamination has been in the soil for a number of years.

APPENDIX

Quality Assurance/Quality Control Procedures

1. Steel probes were used only once during the day and then washed
   with high pressure soap and hot water spray or steam-cleaned to
   eliminate the possibility of cross-contamination. Enough probes
   were carried on the van to avoid the need to reuse any during the
   day.

2. Probe adapters were used to connect the sample probe to the vacuum pump. The adaptor was designed to eliminate the possibility of exposing the soil gas stream to any part of the adaptor. Associated tubing connecting the adaptor to the vacuum pump was replaced periodically as needed during the job to insure cleanliness and good fit.
3. Silicone tubing (which acts as a septum for the syringe needle) was replaced as needed to insure proper sealing around the syringe needle. The tubing does not directly contact soil gas samples.
4. Glass syringes were used for one sample only per day and were washed and baked out at night.
5. Injector port septa through which samples were injected into the chromatograph were replaced on a daily basis to prevent possible gas leaks from the chromatographic column.
6. Analytical instruments were calibrated each day by analytical standards from Chem Service, Inc. Calibration checks were also run after approximately every five sampling locations.
7. Sub-sampling syringes were checked for contamination prior to sampling each day by injecting nitrogen into the gas chromatograph.
8. Prior to sampling each day, system blanks were run to check the sampling apparatus (probe, adaptor, and 10 cc syringe) for contamination by drawing ambient air from above ground through the system and comparing the analysis to a concurrently sampled ambient air analysis.
9. All sampling and sub-sampling syringes were decontaminated each day and no such equipment was reused before being decontaminated. Microliter size sub-sampling syringes were reused only after a nitrogen blank was run to insure it was not contaminated by the previous sample.
10. Soil gas pumping was monitored by a vacuum gauge to insure that an adequate gas flow from the vadose zone was maintained. A reliable gas sample can be obtained if the negative pressure reading on the vacuum gauge was at least 2 inches Hg less than the maximum pressure of the pump.

REFERENCES

[1] Thompson, Glenn M., Favero, Martin D., Golding, Randy D., 1991 "Comparison of Field Headspace Versus Field Soil Gas Analysis Versus Standard Method Analysis of Volatile Petroleum Hydrocarbons in Water and Soil", Proceedings of the Second International Symposium; Field Screening Methods for Hazardous Waste and Toxic Chemicals, Las Vegas, NV, pp. 395-406.

[2] Marrin, Donn L. and Thompson, Glenn M., 1984, "Remote Detection of Volatile Organic Contaminants Via Shallow Soil Gas Sampling" Proceedings of the 1984 Petroleum Hydrocarbons and Organic Chemicals in Groundwater Conference, Houston, Texas: Nat. Water Well Asso., pp. 172-187.

[3] Lappala, Eric G. and Thompson, Glenn M., 1983, "Detection of Groundwater Contamination by Shallow Soil Gas Sampling in the Vadose Zone Theory and Applications", Screening Techniques. p. 20-28.

[4] Marrin, Donn L., 1985, "Delineation of Gasoline Hydrocarbons in Groundwater by Soil Gas Analysis", Proceedings of the 1985 Hazardous Materials Management Conference/West, Long Beach, CA, Dec. 3-5, 1985: Tower Conf. Management Co., pp. 112-119.

[5] Weeks, E.P., et al, 1982, "Use of Atmospheric Fluorocarbons F-11 and F-12 to Determine the Diffusion parameters of the Unsaturated Zone in the Southern High Plains of Texas: Water Resources Research", v. 18, n. 5, Oct. 1982, pp. 1365-1378.

[6] Marrin, Donn L., 1986, "Differences in the Remote Detection of Soil and Groundwater Contamination Using Soil Gas Analysis", Proceedings of the 22nd Symposium on Engineering Geology & Soils Engineering, Boise, ID., Feb. 24-26, 1986, pp. 622-431.

Chi-Yuan Fan[1] and Anthony N. Tafuri[1]

SCREENING METHODOLOGY FOR SELECTING CLEAN-UP TECHNOLOGIES AT LEAKING UNDERGROUND STORAGE TANK SITES

REFERENCE: Fan, C.-Y. and Tafuri, A. N., "Screening Methodology for Selecting Clean-Up Technologies at Leaking Underground Storage Tank Sites," Leak Detection for Underground Storage Tanks, ASTM STP 1161, Philip B. Durgin and Thomas M. Young, Eds., American Society for Testing and Materials, Philadelphia, 1993.

ABSTRACT: This paper presents a methodology for evaluating site conditions and for screening various remediation technologies that may be applicable to cleaning up releases of petroleum products from underground storage tanks (UST). The methodology enables the user to develop a conceptual understanding of site conditions before the completion of extensive field studies, define remediation goals, evaluate technologies capable of meeting remediation goals, and identify monitoring requirements during and after remediation.

Working with site-assessment data, the methodology provides an approach for a preliminary determination of petroleum product location in unsaturated and saturated zones and for evaluating the likelihood of pollutant migration within a site's soil matrix. Worksheets then guide the evaluation of how site conditions pertain to factors that favor or inhibit the success of specific technologies.

KEYWORDS: biorestoration, groundwater, hydraulic barriers, rainfall, soil flushing, soil sorption, soil vapor extraction, underground storage tank, unsaturated zone

Cleaning up releases of petroleum hydrocarbons or other organic chemicals in the subsurface environment is a real-world problem. Treatment of contaminated soil is a major concern, regardless of the location of unsaturated or saturated zones. The science associated with soil treatment technologies, however, is not as well documented or advanced as that associated with water treatment technologies. Despite this uncertainty, soil contamination must always be addressed by engineers on behalf of the tank owners and operators, and some kind of soil cleanup is typically necessary when a release occurs.

In reality, cleanup of the saturated and unsaturated zones should not be considered as unrelated activities. To have an effective overall remediation plan, the potential impact that cleanup in one zone has on another should be considered. Furthermore, the proposed approach for cleaning up the contaminated soil or soil and groundwater must be approved by the regulatory agencies charged with monitoring the cleanup.

[1]Environmental Engineer and Chief, respectively, Releases Technology Section, Risk Reduction Engineering Laboratory, U.S. Environmental Protection Agency, 2890 Woodbridge Avenue, Edison, New Jersey 08837.

Given the uncertainty surrounding the removal of petroleum hydrocarbons from soil, how can one select an effective treatment technology? This paper provides an approach for evaluating the effectiveness of remedial technologies in the face of incomplete or uncertain data. The approach builds on a conceptual understanding of the site conditions and the present state of knowledge of remedial technologies. The methodology is based on two recent EPA publications, one addressing the unsaturated zone [1], the other addressing the saturated zone [2].

Several limitations should be noted when using the approach proposed herein:

- Not a Design Manual - The methodology is not intended to be used as a design manual for engineers whose primary goal is final technology selection and design. Rather, it is intended as a pre-selection methodology, more suited to a preliminary screening of alternatives likely to be effective.

- Not for Emergency Response - The comprehensive cleanup of an UST site involves several stages of remediation. First among these are emergency measures involving tank removal and response to any potentially explosive conditions resulting from the leak. It is assumed that all necessary emergency responses have been taken, that the source of the release has been identified and repaired, and that proper notification of government agencies has taken place.

- Scientific and Engineering Basis for Evaluation - Evaluation of the various cleanup technologies presented is based primarily on a scientific understanding of the processes which govern the mobility of organic contaminants (i.e., move into and out of other phases) in the subsurface environment. The key parameters that control the process kinetics are used as the engineering-related factors to determine the effectiveness of a given technology in cleaning up a contaminated area. Factors for soil venting technology, for example, include the volatility of the contaminants (vapor pressure), soil air conductivity, soil water content, soil sorption capacity, etc. Conditions at any site can be compared with these factors to determine if soil venting is likely to be feasible

  Other considerations, such as site conditions, construction and operation costs, or equipment availability, are often very important in assessing and selecting a technology or a treatment train. The user should be aware of such considerations when assessing a site.

- Focus on Petroleum Hydrocarbons as Contaminants - Assumptions about released contaminants in this paper focus on petroleum products, given the predominance of these materials as stored product in USTs.

- Qualitative Cleanup Goals - Technology selection is usually based on what cleanup goals must be achieved. This paper only addresses a qualitative discussion of cleanup efficiency.

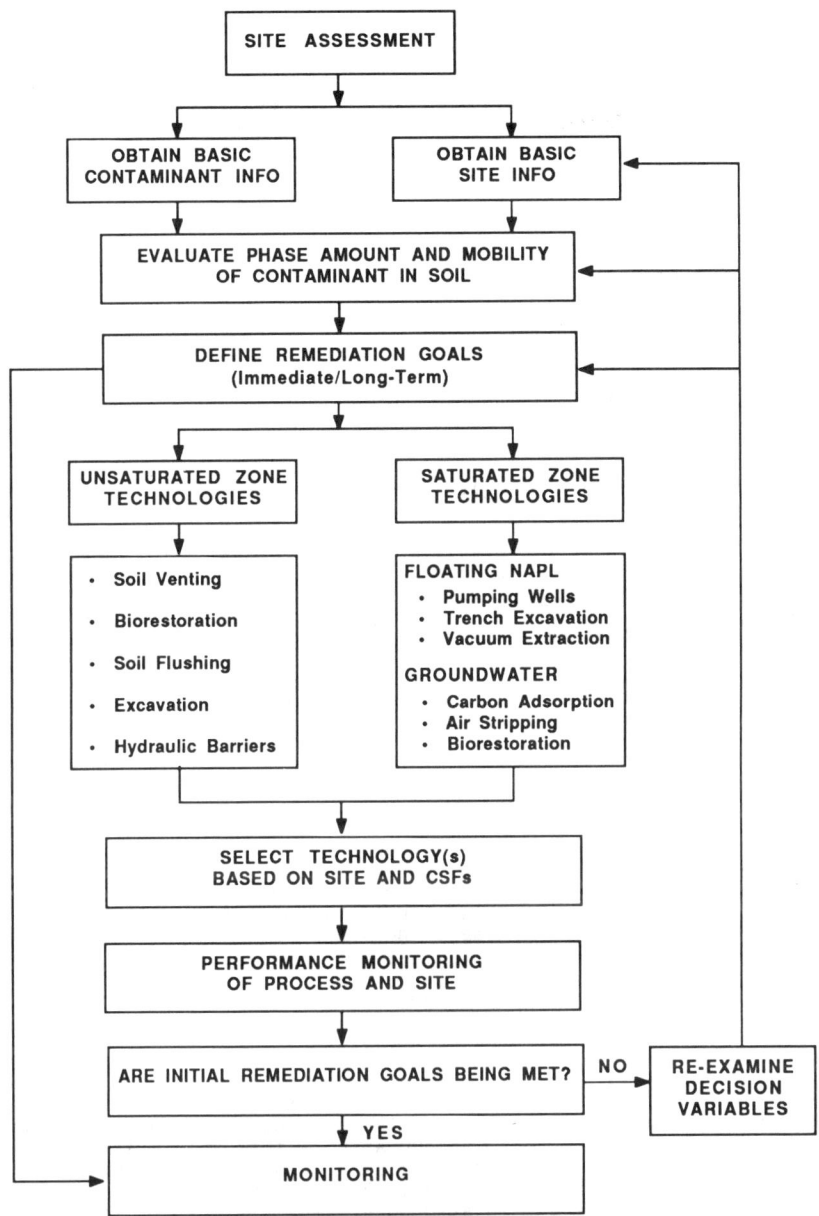

FIGURE 1--Overview of Technology Screening Methodology

OVERVIEW OF APPROACH

An effective response to underground storage tank releases requires understanding site conditions, defining appropriate remediation goals, selecting an effective clean-up and treatment system as well as the monitoring and follow-up measurements. Figure 1 shows an overview of the selection process.

SITE ASSESSMENT

A site assessment begins with basic information about the release itself. Questions are:

- What contaminants were released?
- How much was released?
- Was the release slow or instantaneous?
- How long ago did the release stop?
- When was the release detected?

Information about the site is gathered next. This usually involves searches of records, use of professional expertise, and, if necessary, sources of default values for making ballpark estimates. Typical site information includes:

- Soil temperature
- Soil surface area
- Particle density
- Depth to groundwater
- Infiltration rate

- Soil pH
- Soil porosity
- Permeability
- Field capacity
- Rainfall intensity

- Soil organic content
- Bulk density
- Soil moisture content
- Hydraulic conductivity
- Water Quality

A site investigation should also include an understanding of the physical and chemical nature of the contaminants released. With this, it is possible to estimate how the contaminant may partition in the subsurface (what phase it is likely to be in), how readily it will move away from the site as a vapor, liquid, or dissolved into groundwater. Contaminant-specific parameters include:

- Unweathered composition
- Soil sorption coefficient
- Water solubility
- Density of non-aqueous phase liquids (NAPL)
- Viscosity of non-aqueous phase liquids

- Pore liquid pressure
- Pore vapor pressure
- Vapor density

A recent EPA study indicates that petroleum hydrocarbons can be found in 13 different phases-locations in the subsurface environment [3]. For simplicity, this writeup assumes that they exist in four phases (vapor, NAPL, adsorbed on soil, and dissolved in pore water) in the vadose zone and three phases (NAPL, adsorbed on soil, and dissolved in groundwater) in the saturated zone. Mobility, a critical factor in determining remediation goals, is used here as a general term to indicate how readily a contaminant moves into air and water in the subsurface environment.

The primary indicators in determining mobility are vapor pressure (how easily will the contaminant volatilize into soil pore space) and solubility (what is the contaminant's affinity for water). For a recent continuous release, most of the petroleum products are likely to be in the nonaqueous liquid phase, with somewhat smaller portions existing in the vapor and dissolved phase.

Tables 1 through 3 present worksheets that indicate the factors that may influence a vapor or liquid's mobility in the subsurface environment. If the preponderance of factors at a site fall in the right-hand columns of the worksheet, contaminants would probably be more mobile and likely to migrate than if most of the factors match those in the left-hand column.

DEFINE REMEDIATION GOALS

A remediation plan must address short-term and long-term cleanup goals. Complete restoration at a release site is desirable, but it is not economically feasible to achieve pristine conditions at all sites. Therefore, the remediation goals should be based upon the unique conditions at each site and be realistically determined for each cleanup phase (i.e. source removal for emergency response, vadose zone cleanup, and groundwater treatment).

The time-frame over which each phase of remediation is implemented will vary from site to site. Emergency measures should always be addressed and implemented quickly to protect immediate health effects. Then, a complete long-term remediation plan should be developed based on the prioritization of the most urgent or important tasks. For example, a containment system to prevent further contaminant migration may take precedence over treatment of soil and groundwater. Initial design of a containment system should focus on preventing migration of NAPL and contaminated groundwater. However, in some cases, a combination of containment and treatment of groundwater may be a long-term objective of remediation, if treatment alone is not considered sufficient.

SCREENING TECHNOLOGIES

The selection of cleanup technologies can be put into three broad categories: 1) remediation of the unsaturated zone, 2) removal of nonaqueous phase liquid, and 3) treatment of contaminated groundwater. For each cleanup technology, factors critical to the effectiveness of each technology are identified in the worksheets shown in Tables 4 through 14. Many of the parameters determined during the site assessment serve as critical success factors (CSFs) of how well a given cleanup technology will perform.

The worksheets contain CSFs for screening the likelihood of success of several technologies at a particular site. Conditions at any site can be compared with a treatment technology's CSFs to determine its effectiveness. CSFs are not equally important. Depending on site conditions, some CSFs may be much more important than others. One critical factor may override all others. For instance, the most important factor for the soil vapor extraction technology is contaminant vapor pressure closely followed by soil permeability. Users must make their own judgment to prioritize the CSFs for specific conditions at the site. A brief description of each of the technologies and the related CSFs is provided below.

Remediation of unsaturated zone

Five technologies commonly used to remediate vadose zone contamination are soil venting, biorestoration, soil flushing, hydraulic barriers, and excavation. The first four are in situ treatment methods,

TABLE 1—*Factors to evaluate the mobility of liquid contaminants.*

SITE:					
FACTOR	**UNITS**	**DATA**	**INCREASING MOBILITY →**		
RELEASE-RELATED INFORMATION					
Time Since Release	Months		Long (>12) ☐	Medium (1-12) ☐	Short (<1) ☐
SITE-RELATED INFORMATION					
Hydraulic Conductivity	cm/sec		Low (<10^{-5}) ☐	Med. (10^{-5} - 10^{-3}) ☐	High (>10^{-3}) ☐
Soil Porosity	% Soil Volume		High (>40) ☐	Medium (20-30) ☐	Low (<30) ☐
Soil Surface Area	m²/g		High (>1) ☐	Medium (0.1- 1) ☐	Low (<0.1) ☐
Soil Temperature	°C		Low (<10) ☐	Medium (10-20) ☐	High (>20) ☐
Rock Fractures	...		Absent ☐	...	Present ☐
Moisture Content	% Volume		High (>30) ☐	Medium (10-30) ☐	Low (<10) ☐
CONTAMINANT-RELATED INFORMATION					
Liquid Viscosity	cPoise		High (>20) ☐	Medium (2-20) ☐	Low (<2) ☐
Liquid Density	g/cm3		Low (<1) ☐	Medium (1-2) ☐	High (>2) ☐

TABLE 2--*Factors to evaluate the mobility of contaminant vapors.*

SITE:					
FACTOR	UNITS	DATA	INCREASING MOBILITY→		
SITE-RELATED INFORMATION					
Air Filled Porosity	% Volume		High (>40) ☐	Medium (20-30) ☐	Low (<30) ☐
Total Porosity	% Volume		High (>40) ☐	Medium (20-30) ☐	Low (<30) ☐
Water Content	% Volume		High (>30) ☐	Medium (10-30) ☐	Low (<10) ☐
Depth Below Surface	Meters		Deep (>10) ☐	Medium (2-10) ☐	Shallow (<2) ☐
CONTAMINANT-RELATED INFORMATION					
Vapor Density	g/m3		Low (<50) ☐	Medium (50-500) ☐	High (>500) ☐

TABLE 3--*Factors to evaluate the mobility of contaminants in pore water.*

SITE:					
FACTOR	UNITS	DATA	INCREASING MOBILITY→		
SITE-RELATED INFORMATION					
Hydraulic Conductivity	cm/sec		Low (<10^{-5}) ☐	Med.(10^{-5} - 10^{-3}) ☐	High (>10^{-3}) ☐
Moisture Content	% Volume		Low (<10) ☐	Medium (10-30) ☐	High (>30) ☐
Rainfall Infiltration Rate	cm/day		Low (<0.05) ☐	Medium (0.05-0.1) ☐	High (>0.1)) ☐
Soil Porosity	% Volume		High (>40) ☐	Medium (20-30) ☐	Low (<30) ☐
Rock Fractures	...		Absent ☐	...	Present ☐
Depth Below Surface	meters		Shallow (<2) ☐	Medium (2-10) ☐	Deep (>10) ☐
Water Solubility	mg/L		Low (<100) ☐	Med. (100-1000) ☐	High (>1000) ☐

and the fifth one is ex-situ, aboveground either on site or at a centralized treatment/disposal facility. The five methods and factors critical to the success of each technology are presented in Tables 4 through 8.

Soil venting--Soil venting (Table 4) is a general term that refers to any technique that removes volatile organic vapors from the vadose zone. It may be passive or active. For removing petroleum hydrocarbon vapor in the vadose zone at a leaking underground storage site, active venting is the most effective and widely used process. Active venting uses extraction wells to induce negative pressure gradient to move vapors through the soil layer toward the well; thus, it is commonly called vacuum extraction or soil vapor extraction (SVE) technology. The success of a soil vapor extraction system depends both on the properties of the contaminants and on the properties of the soil. The volatility of a contaminant is an important characteristic affecting the applicability of SVE to that compound. The parameter that best describes a compound's volatility is its vapor pressure.

Biorestoration--In situ biorestoration (Table 5) of the unsaturated zone is a process where oxygen and sometimes nutrients are added to contaminated soil to promote the breakdown of petroleum hydrocarbons by naturally occurring microorganisms to carbon dioxide, water, and other non-toxic end products. For a successful biorestoration system, an adequate supply of oxygen and nutrients must be available to the microorganisms throughout the zone of contamination. This is usually accomplished by adding nutrient-enhanced water to the vadose zone through infiltration basins or through recharge wells. Oxygen may be introduced to the subsurface by injecting either air or hydrogen peroxide.

A biorestoration system can be installed relatively quickly, but it may take several months for the microbes to start significant degradation, particularly if the contaminant release is recent. Furthermore, the system may require periodic adjustment of the dosages of oxygen and nutrient feeds in order to operate efficiently.

Soil flushing--Soil flushing (Table 6) refers to an in situ process that involves flooding the zone of contamination with water or a water-surfactant mixture in order to dissolve the contaminants into the water. The contaminated water is then pumped above to the groundwater treatment facility by extraction wells. The well network must be strategically designed and installed so that the mobilized mixture of contaminants and water will not escape to the saturated zone to cause groundwater contamination. A compound's soil/water partitioning coefficient and water solubility are the most important factors.

Hydraulic methods--Hydraulic methods (Table 7) include sumps, French drains, and other designs to create an area of high permeability in the vadose zone for collecting and removing the accumulated mobile nonaqueous phase liquid and contaminated soil water. The methods are simple and relatively inexpensive, but do not remove the residual portion of the contaminants. Thus, they are likely to be selected as an initial step in the cleanup option. Other physical barriers to flow (e.g. caps, slurry walls, etc.) are not addressed here because they are typically not practical at leaking underground storage tank sites.

Excavation--As an alternative to the four in situ treatment methods, excavated soil may be treated on site or treated and disposed at an off- site facility (Table 8). On-site treated soil is sometimes placed back into the excavation area. At present, excavating contaminated soil is more common than in situ treatment. Excavation is easy to undertake and may be done quickly (depending upon the depth of contamination and soil type), and it has the ability to remove most or

TABLE 4—Worksheet for evaluating the feasibility of soil venting.

SITE:

CRITICAL SUCCESS FACTOR	UNITS	DATA	SUCCESS LESS LIKELY	SUCCESS SOMEWHAT LIKELY	SUCCESS MORE LIKELY
SITE-RELATED INFORMATION					
Dominant Contaminant Phase	Phase		Sorbed to soil ☐	Liquid ☐	Vapor ☐
Soil Temperature	°C		Low (<10) ☐	Medium (10-20) ☐	High (>20) ☐
Soil Air Conductivity	cm/sec		Low (<10^{-6}) ☐	Medium (10^{-6}-10^{-4}) ☐	High (>10^{-4}) ☐
Moisture Content	% Volume		High (>30) ☐	Medium (10-30) ☐	Low (<10) ☐
Geological Conditions	...		Heterogeneous	...	Homogeneous
Soil Sorption Capacity Surface Area	m^2/g		High (>1) ☐	Medium (0.1-1) ☐	Low (<0.1) ☐
Depth to Groundwater	meters		Low (<1) ☐	Medium (1-5) ☐	High (>5) ☐
CONTAMINANT-RELATED INFORMATION					
Vapor Pressure	mm Hg		Low (<10) ☐	Medium (10-100) ☐	High (>100) ☐
Water Solubility	mg/L		High (>1000) ☐	Med. (100-1000) ☐	Low (<100) ☐

TABLE 5—Worksheet for evaluating the feasibility of biorestoration.

SITE:

CRITICAL SUCCESS FACTOR	UNITS	DATA	SUCCESS LESS LIKELY	SUCCESS SOMEWHAT LIKELY	SUCCESS MORE LIKELY
RELEASE-RELATED INFORMATION					
Time Since Release	Months		Short (<1) ☐	Medium (1-12) ☐	Long (>12) ☐
SITE-RELATED INFORMATION					
Dominant Contaminant Phase	Phase		Liquid ☐	Vapor ☐	Dissolved ☐
Soil Temperature	°C		Low (<5) ☐	Medium (5-10) ☐	High (>10) ☐
Soil Hydraulic Conductivity	cm/sec		Low (<10^{-5}) ☐	Medium (10^{-5}-10^{-3}) ☐	High (<10^{-3}) ☐
Soil pH	pH units		(<6 or <8) ☐	...	(6-8) ☐
Moisture Content Volume	% Volume		Dry (<10) ☐	Moderate (10-30) ☐	Moist (>30) ☐
CONTAMINANT-RELATED INFORMATION					
Solubility	mg/L		Low (<100) ☐	Med (100-1000) ☐	High (>1000) ☐
Biodegradability Refractory Index	...		Low (<0.01) ☐	Medium (0.01-0.1) ☐	High (>0.1) ☐
Fuel Type			No. 6 Fuel Oil (heavy) ☐	No. 2 Fuel Oil (medium) ☐	Gasoline/ Diesel (light) ☐

TABLE 6—Worksheet for evaluating the feasibility of soil flushing.

SITE:

CRITICAL SUCCESS FACTOR	UNITS	DATA	SUCCESS LESS LIKELY	SUCCESS SOMEWHAT LIKELY	SUCCESS MORE LIKELY
SITE-RELATED INFORMATION					
Dominant Contaminant Phase	Phase		Vapor ☐	Liquid ☐	Dissolved ☐
Soil Hydraulic Conductivity	cm/sec		Low ($<10^{-5}$) ☐	Medium (10^{-5}-10^{-3}) ☐	High ($>10^{-3}$) ☐
Soil Surface Area	m²/g		High (>1) ☐	Medium (0.1-1) ☐	Small (>0.1) ☐
Carbon Content	% Weight		High ($>10\%$) ☐	Medium (1-10%) ☐	Low ($<1\%$) ☐
Fractures in Rock	...		Present ☐	...	Absent ☐
CONTAMINANT-RELATED INFORMATION					
Water Solubility	mg/L		Low (<100) ☐	Med (100-1000) ☐	High (>1000) ☐
Sorption Characteristics Soil Sorption Constant	L/kg		High ($>10,000$) ☐	Medium (100-10,000) ☐	Low (<100) ☐
Vapor Pressure	mm Hg		High (>100) ☐	Medium (10-100) ☐	Low (<10) ☐
Liquid Viscosity	cPoise		High (>20) ☐	Medium (2-20) ☐	Low (<2) ☐
Liquid Density	g/cm³		Low (<1) ☐	Medium (1-2) ☐	High (>2) ☐

TABLE 7—Worksheet for evaluating the feasibility of hydraulic barriers.

SITE:

CRITICAL SUCCESS FACTOR	UNITS	DATA	SUCCESS LESS LIKELY	SUCCESS SOMEWHAT LIKELY	SUCCESS MORE LIKELY
RELEASE-RELATED INFORMATION					
Time Since Release	Months		Long (>12) ☐	Medium (1-12) ☐	Short (<1) ☐
Volume of Spill	Liters		Small (<400) ☐	Medium (400-4000) ☐	Large (>4000) ☐
SITE-RELATED INFORMATION					
Dominant Contaminant Phase	Phase		Vapor ☐	...	Liquid ☐
Soil Hydraulic Conductivity	cm/sec		High ($>10^{-3}$) ☐	Medium (10^{-5}-10^{-3}) ☐	Low ($<10^{-5}$) ☐
Soil Sorption Capacity Surface Area	m²/g		High (>1) ☐	Medium (0.1-1) ☐	Low (<0.1) ☐
Carbon Content	% Weight		High ($>10\%$) ☐	Medium (1-10%) ☐	Low ($<1\%$) ☐
Temperature	°C		Low (<5) ☐	Medium (5-10) ☐	High (>10) ☐
Depth to Groundwater	Meters		High (>5) ☐	Medium (1-5) ☐	Low (<1) ☐
CONTAMINANT-RELATED INFORMATION					
Liquid Viscosity	cPoise		High (>20) ☐	Medium (2-20) ☐	Low (<2) ☐

all of the contamination from the site to an off-site treatment/disposal facility. A wide variety of above-ground processes may be used for treating excavated soil, including low temperature thermal desorption, incineration, soil washing, biodegradation (i.e., composting and slurry reactor), using soil as an aggregate in asphalt production, and others.

Recovery or containment of floating NAPL

Technologies for recovery or containment of floating NAPL are pumping wells and trench excavation. NAPL recovery can also be accomplished by vacuum extraction (or vacuum-enhanced pumping), which is a specialized form of soil venting. It has not, however, been widely applied at UST sites. Worksheets for each of these technologies are illustrated in Tables 9, 10, and 11.

Pumping wells--The two main types of recovery systems involving pumping wells are single pump and dual pump systems (Table 9). In single pump systems, one pump is used to remove both groundwater and NAPL. For dual systems, one pump is used to create a depression in the water table and another to remove floating NAPL. Dual pump systems are more commonly used than single pump methods, because mixing of groundwater and NAPL is avoided. Also, the NAPL recovery pump need only be operating when a significant layer of NAPL is accumulated. Above-ground facilities for water treatment and NAPL storage should be provided.

Trench excavation--Trench excavation (Table 10) is effective for recovery and containment of floating NAPL. The liquid phase will move in the direction of the groundwater. Therefore, it is installed downgradient of the plume to intercept the floating liquid and to prevent further migration of the contaminants. Oil skimmers and oil separators are usually equipped to remove floating NAPL in the trench.

Vacuum extraction--Vacuum extraction (or vacuum-enhanced pumping) (Table 11) with or without air sparging has been shown to improve NAPL removal in the capillary fringe zone. The process will increase removal of the mobile portion of the liquid phase and enhance the volatilization of the residual NAPL by inducing air into the saturated zone. One advantage of using this method is that the system can be designed to treat contaminants in both the vadose zone and the saturated zone.

Remediation of dissolved phase contaminants in saturated zone

The two basic types of technologies for remediating groundwater with dissolved contaminants are in situ biorestoration and groundwater pump-and-treat methods. Tables 12, 13, and 14 present worksheets for evaluating the feasibility of using these technologies to treat groundwater.

In Situ biorestoration--In situ biorestoration (Table 12) of the contaminated groundwater has been developed and applied more commonly than remediation of soil in the unsaturated zone. The system includes: injection wells to introduce oxygen-enriched water with or without nutrients upstream of the contaminated plume in the saturated zone; downstream recovery wells to prevent spreading of the dissolved contaminants; and, above-ground nutrients, oxygen mixing and feeding equipment. Water withdrawn through the recovery well can be recirculated through the chemical feeding facility and injection wells back to the aquifers.

Groundwater pump-and-treat systems--Pump-and-treat technology involves extracting the groundwater and treating it at an above-ground

TABLE 8—Worksheet for evaluating the feasibility of excavation.

SITE:

CRITICAL SUCCESS FACTOR	UNITS	DATA	SUCCESS LESS LIKELY	SUCCESS SOMEWHAT LIKELY	SUCCESS MORE LIKELY
RELEASE-RELATED INFORMATION					
Proximity of Above and Below Ground Structures	...		Buildings nearby buried pipes and cables ☐	...	No nearby structures ☐
Volume of Soil Contaminated	Cubic Meters		Large (>1000) ☐	Medium (100-1000) ☐	Small (<100) ☐
Depth of Contamination	Meters from Surface		Deep (>5) ☐	Medium (1-5) ☐	Shallow (<1) ☐
Proximity of Site to Traffic	...		Near ☐	...	Far ☐
Businesses	...		Near ☐	...	Far ☐
Disposal Site	...		Far ☐	...	Near ☐
Backfill Source	...		Far ☐	...	Near ☐

TABLE 9—Worksheet for evaluating the feasibility of using pumping wells to remove floating NAPL.

SITE:

CRITICAL SUCCESS FACTOR	UNITS	DATA	SUCCESS LESS LIKELY	SUCCESS SOMEWHAT LIKELY	SUCCESS MORE LIKELY
RELEASE-RELATED INFORMATION					
Amount Released	Liters		Small (<200,000) ☐	Medium (200,000-2,000,000) ☐	Large (>2,000,000) ☐
Time Since Released	Months		Short (<1) ☐	Medium (1-12) ☐	Long (>12) ☐
SITE-RELEASED INFORMATION					
Site Stratigraphy	...		Complex ☐	...	Simple ☐
Depth to Ground Water	Meters		Shallow (<5) ☐	...	Deep (>5) ☐
CONTAMINANT-RELATED INFORMATION					
Liquid Density	g/cm³		Low (<1) ☐	...	High (>1) ☐
Liquid Viscosity	cP		High (>2) ☐	Medium (1-2) ☐	Low (<1) ☐

TABLE 10--**Worksheet for evaluating the feasibility of using trench excavation to contain floating NAPL.**

SITE:					
CRITICAL SUCCESS FACTOR	UNITS	DATA	SUCCESS LESS LIKELY	SUCCESS SOMEWHAT LIKELY	SUCCESS MORE LIKELY
RELEASE-RELATED INFORMATION					
Amount Released	Liters		Large (>2,000,000) ☐	Medium (200,000-2,000,000) ☐	Small (<200,000) ☐
Time Since Released	Months		Long (>12) ☐	Medium (1-12) ☐	Short (<1) ☐
SITE-RELEASED INFORMATION					
Depth to Ground Water	Meters		Deep (>5) ☐	Medium (1-5) ☐	Shallow (<1) ☐
Stability of Soil Formation	...		Unstable ☐	...	Stable ☐
Presence of Interfering Structures	...		Present ☐	...	Not Present ☐
CONTAMINANT-RELATED INFORMATION					
Liquid Viscosity	cP		High (>2) ☐	Medium (1-2) ☐	Low (<1) ☐
Liquid Density	g/cm^3		High (>1) ☐	...	Low (<1) ☐

TABLE 11—Worksheet for evaluating the feasibility of using vacuum extraction to remove floating NAPL.

SITE: _____

CRITICAL SUCCESS FACTOR	UNITS	DATA	SUCCESS LESS LIKELY	SUCCESS SOMEWHAT LIKELY	SUCCESS MORE LIKELY
SITE-RELATED INFORMATION					
Soil Air Conductivity	cm/sec		Low (<10⁻⁶) ☐	Medium (10⁻⁴-10⁻⁶) ☐	High (>10⁻⁴) ☐
Soil Temperature	°C		Low (<10) ☐	Medium (10-20) ☐	High (>20) ☐
Moisture Content	% Volume		High (>30) ☐	Medium (10-30) ☐	Low (<10) ☐
Soil Surface Area	m²/g		High (>1) ☐	Medium (0.1-1) ☐	Low (<0.1) ☐
Carbon Content	% Weight		High (>10) ☐	Medium (1-10) ☐	Low (<1) ☐
CONTAMINANT-RELATED INFORMATION					
Vapor Pressure	mm Hg		Low (<10) ☐	Medium (10-100) ☐	High (>100) ☐
Water Solubility	mg/L		High (>1000) ☐	Medium (100-1000) ☐	Low (<100) ☐

TABLE 12—Worksheet for evaluating the feasibility of using biorestoration to treat groundwater.

SITE: _____

CRITICAL SUCCESS FACTOR	UNITS	DATA	SUCCESS LESS LIKELY	SUCCESS SOMEWHAT LIKELY	SUCCESS MORE LIKELY
RELEASE-RELATED INFORMATION					
Time Since Release	Months		Short (<1) ☐	Medium (1-12) ☐	Long (>12) ☐
SITE-RELATED INFORMATION					
Hydraulic Conductivity	cm/sec		Low (<10⁻⁵) ☐	Medium (10⁻⁵-10⁻³) ☐	High (<10⁻³) ☐
Site Stratigraphy	...		Complex ☐	...	Simple ☐
Groundwater Temperature	°C		Low (<5) ☐	Medium (5-10) ☐	High (>10) ☐
pH	pH Units		(<6 or >8) ☐	...	(6-8) ☐
CONTAMINANT-RELATED INFORMATION					
Biodegradability Refractory Index	...		Low (<0.01) ☐	Medium (0.01-0.1) ☐	High (>0.1) ☐
Total Organic Carbon Content of GW	mg/L		(<10 or >1000) ☐	...	(10-1000) ☐

treatment facility. This method is appropriate for most groundwater contamination problems, especially for petroleum hydrocarbons that are lighter than water. However, the technology may not be efficient for heterogeneous aquifer conditions where low hydraulic conductivity zones restrict contaminant flow toward the extraction wells. The system includes a set of extraction wells, pumping equipment, and water treatment processes. The two most commonly used treatment process units for removing dissolved petroleum hydrocarbons are air stripping (Table 13) and activated carbon adsorption (Table 14). Of the air stripping methods, packed towers work best at removing VOCs from groundwater. However, the quality of effluent from a properly designed carbon adsorption unit can achieve drinking water standards.

Air sparging--Air sparging is an innovative technology which is being used at an increasing number of sites for the remediation of contamination in the saturated zone. The system involves the injection of oil-free air into the saturated zone to strip volatile organic chemicals dissolved in groundwater and adsorbed to soils from the saturated zone. The vapor phase contaminants transferred to the unsaturated zone are then captured using soil vapor extraction. Understanding the characteristics of air transport in both the saturated and unsaturated zones is important to evaluate the effectiveness of air sparging systems.

MONITORING

The purpose of monitoring is to determine the amount and movement of pollutants in the subsurface environment before, during, and after remediation. An effective monitoring program includes the design of a reliable well network to ensure a complete determination of and assessment of the site conditions. The following should be considered in setting up an effective monitoring network:

- At least one well should be placed up-gradient or outside of the zone of contamination produced by the release so that accurate background levels can be monitored.

- There should be enough down-gradient wells to adequately monitor the horizontal and vertical extent of contamination, especially in complex stratigraphy or fluctuating ground water table areas.

- Wells should be screened in the zone of contamination at appropriate depths. Special attention should be given to the design of the length of well screens. Longer screens are more likely to intercept the contaminant plume, but may result in diluted soil gas or ground water samples. Shorter screens provide better concentration estimates but require more accurate location to ensure the plume is intercepted.

- Well screen length in the capillary zone must be longer than the total depth of ground water fluctuations in order to adequately monitor floating NAPL.

- All pathways for potential migration of contaminants should be monitored in multiple stratigraphies and aquifers.

The overall objectives of a monitoring program are to: (1) assess the site conditions to determine a remediation approach, including the feasibility and requirement of a "no action" decision; (2) evaluate the progress of in situ treatment; and (3) determine site conditions after treatment is terminated.

TABLE 13—Worksheet for evaluating the feasibility of using air stripping to treat extracted groundwater.

SITE:

CRITICAL SUCCESS FACTOR	UNITS	DATA	SUCCESS LESS LIKELY	SUCCESS SOMEWHAT LIKELY	SUCCESS MORE LIKELY
RELEASE-RELATED INFORMATION					
Amount Release	Liters		High (<4,000) ☐	Medium (4,000-200,000) ☐	Large (>200,000) ☐
Time Since Release	Months		Short (<1) ☐	Medium (1-12) ☐	Long (>12) ☐
SITE-RELATED INFORMATION					
Groundwater Temperature	°C		Low (<10) ☐	Medium (10-20) ☐	High (>20) ☐
Total Suspended Solids Content of GW	mg/L		High (>20) ☐	Medium (5-20) ☐	Low (<5) ☐
Total Dissolved Iron and Manganese Content of GW	mg/L		High (>5) ☐	Medium (0.2-5) ☐	Low (<0.2) ☐
CONTAMINANT-RELATED INFORMATION					
Vapor Pressure	mm Hg		Low (<10) ☐	Medium (10-100) ☐	High (>100) ☐
Water Solubility	mg/L		High (>1000) ☐	Medium (100-1000) ☐	Low (<100) ☐
Dissolved Contaminant Concentration	mg/L		Low (<1) ☐	Medium (1-100) ☐	High (>100) ☐

TABLE 14—Worksheet for evaluating the feasibility of using carbon adsorption to treat extracted groundwater.

SITE:

CRITICAL SUCCESS FACTOR	UNITS	DATA	SUCCESS LESS LIKELY	SUCCESS SOMEWHAT LIKELY	SUCCESS MORE LIKELY
RELEASE-RELATED INFORMATION					
Amount Release	Liters		Large (>2,000,000) ☐	Medium (4,000-200,000) ☐	Small (<4,000) ☐
Time Since Release	Months		Long (>12) ☐	Medium (1-12) ☐	Short (<1) ☐
SITE-RELATED INFORMATION					
TOC Content of GW	mg/L		High (>5) ☐	Medium (1-5) ☐	Low (<1) ☐
Suspended Solids Content of GW	mg/L		High (>20) ☐	Medium (5-20) ☐	Low (<5) ☐
Total Dissolved Iron and Manganese Content of GW	mg/L		High (>5) ☐	Medium (0.2-5) ☐	Low (<0.2) ☐
CONTAMINANT-RELATED INFORMATION					
Water Solubility	mg/L		High (>1000) ☐	Medium (100-1000) ☐	Low (<100) ☐
Molecular Weight	g/mole		Low (<100) ☐	Medium (100-200) ☐	High (>200) ☐

Monitoring to Select Remediation Approach

After completion of the desk-top analysis of site conditions and treatment technology screening, a field monitoring program should be conducted to finalize the selection and design of a corrective action approach. The monitoring program will include field sampling to verify soil and site characteristics and to confirm previous assumptions regarding the subsurface environment. It is also important to properly design and operate a monitoring well network to determine contaminant movement and to examine passive biodegradation potential. (There are some cases where naturally occurring biodegradation may be a feasible option to attain site remediation. To assess this "no action" alternative and to evaluate decision criteria, a properly designed monitoring network in conjunction with site specific vapor and ground water transport modeling, should be performed.)

Monitoring Remediation Progress

Once a technology has been selected, designed and installed, it is essential that the performance of the treatment system be continually evaluated to ensure that it is operating effectively. Remediation of petroleum product releases is often a long process. A treatment technology is usually designed to remove one or more specific constituents to a specified level that is often set to conform to regulatory standards. Performance is then evaluated by measuring the concentrations of each contaminant of concern, and comparing those levels to cleanup goals.

Measuring the progress of floating NAPL removal is usually accomplished simply by comparing the volume of NAPL removed to an estimate of the total floating NAPL. However, measuring removal progress for contaminants dissolved in soil water and/or adsorbed on soil particles is more difficult. An effective in situ monitoring well network is necessary to evaluate how the technology is working and how the overall cleanup is progressing. Proper monitoring can detect problems in the design or implementation of the selected technology. If a technology is performing poorly, it may be due to an improper design arising from an incomplete site characterization. In this case, it may be necessary to re-examine the data that were collected for the site assessment. More data may be needed to enhance the site investigation. Design modifications may be warranted if the understanding of site conditions changes appreciably.

Post Remediation Operation Monitoring

Remediation of a site is complete when the cleanup goals and cleanup criteria are met and maintained. Just meeting the cleanup criteria is not a sufficient reason to suspend remediation. Monitoring of the site should continue after cleanup because contamination levels can increase after treatment stops. The following are some of the causes of increased contamination levels at a site.

- Adsorbed contaminants or contaminants in low permeability zones can persist in the subsurface but may not be detected at monitoring wells while the system is operating. After shutdown these contaminants will tend to disperse causing further increases in contamination levels in soil gas or ground water.

- Soil gas or ground water flow patterns created by extraction wells can dilute samples. After pumping stops, normal flow patterns return and concentration levels may increase.

Therefore, the decision to discontinue monitoring should be made jointly with regulatory officials and experienced professionals to ensure that remediation is actually complete.

REFERENCES

[1] Lyman, W. J., and Noonan, D. C., "Assessing UST Corrective Action Technologies: Site Assessment and Selection of Unsaturated Zone Treatment Technologies," EPA/600/2-90/011, U.S. Environmental Protection Agency, March 1990.

[2] Reidy, P. J., Lyman, W. J., and Noonan, D. C., "Assessing UST Corrective Action Technologies: Early Screening of Cleanup Technologies for the Saturated Zone," EPA/600/2-90/027, U.S. Environmental Protection Agency, June 1990.

[3] Lyman, W. J., Reidy, P. J., and Levy, B., "Assessing UST Corrective Action Technology: A Scientific Evaluation of the Mobility and Degradability of Organic Contaminants in Subsurface Environments," EPA/600/2-91/053, U.S. Environmental Protection Agency, September 1991.

Author Index

D

Durgin, P. B., 115

E

Eckert, E. G., 53

F

Fan, C.-Y., 211
Ferguson, K. W., 189
Fleischer, D. W., 17
Flora, J. D., Jr., 30, 151

G

Glauz, W. D., 30, 151
Golding, R. D., 131, 197
Grace, R., 90
Grey, A. E., 105
Gulledge, W. P., 167
Guzman, A. M., 90

H

Hennon, G. J., 30, 151
Hibner, J., 90
Hillger, R. W., 53, 175

M

MacArthur, M. P., 175
Maresca, J. W., Jr., 53, 175
Martin, M. V., 123
Michelson, R. W., 115

P

Partin, J. K., 105
Portnoff, M. A., 90

R

Rogers, W. F., 3
Rosenberg, M. S., 73

S

Schreiber, R. P., 73
Starr, J. W., 175
Sutton-Mendoza, S. A., 162

T

Tafuri, A. N., 211
Thompson, G. M., 131

W

Wichman, T. A., 197

Y

Yezzi, J. J., 53
Young, T. M., 139

Subject Index

A

Absorption, 123, 131
Acetone, 175
Acoustic emission, 53
Acoustic sensing methods, 53
Air permeability, 131
Airports
　hydrant systems, 30
　leak detection at, 131
Automated gas testing, 90
Automated monitoring systems
　line detectors, 151
　tank gauge, 3, 17, 139

B

Bulk modulus, 151
Butane, 90

C

Certification, 167
Chart errors, tank, 17
Chemical sensing, 105
Chemical tracers, 131
Chromatograph, gas, field mobilized, 197
Cleanup, site, 73, 105, 211
Coefficient of expansion, 17
Compliance, 167, 189

D

Diffusion, vapor, 73, 123

E

Enforcement, state regulatory, 162
Environmental impairment liability insurance, 167
Evaporation, 17
Exposure potential, 189

F

Fiber optic sensors, 105
Field analytical methods, 197
Field inspections, 162
Frozen soil, effects of, 151
Fuel storage tank, 3

G

Gasoline, 115
Gas sensor, 90
Gauge, tank, automatic, 3, 17, 139

H

Hazardous chemicals, 175
Humidity, sensor effects by, 90
Hydrant systems, airport, 30
Hydrocarbons, 73, 115, 197

I

Independent testing, 139
Inspections, field, New Mexico, 162
Installation, monitoring points, 115
Institute for Environmental Auditing, 167
Insurance, environmental impairment liability, 167
Interstitial monitoring, piping, 30
Inventory analysis, 3
Iowa, hydrocarbon contamination, 197

L

Leak rate, 3
Level measurement, 17
Light transmission, 105
Liquid detectors, 139
Line tests, 151
Location, leak, 53

M

Management plans, 189
Metal oxide semiconductor gas sensor, 90
Methane, 90
Methanol, 175
Methyl ethyl ketone, 175
Modeling, 73
    risk assessment, 189
Monitoring system design, 73

N

New Mexico, storage tanks enforcement program, 162

O

Overfill/spill protection, 115

P

Passive acoustics, 53
Permeability
    air, 131
    soil, 197
Petroleum pipes, 53
Pipes, 123, 131
    petroleum, 53
    underground, 17, 30
Pollutant migration, 211
Polymer adsorption gas sensor, 90
Pressure-step methods, 30

R

Rainfall, 211
Regulatory reinforcement, New Mexico, 162
Remediation, 73, 105, 211
Risk assessment, 189

S

Sensors and sensor methods
    acoustic, 53
    fiber optic, 105
    metal oxide semiconductor, 90
    polymer adsorption gas, 90
    tubes, 123

Site assessments, 115, 167, 197, 211
Soil matrix, 211
Soil moisture content, 123, 151
Soil sorption, 211
Soil vapor, 73, 131, 197
    extraction, 211
Standards, 167
Statistical evaluation, 151
    inventory reconciliation methods, 139
Steel tanks, 175
System design, monitoring, 105

T

Tank gauging, automatic, 3, 17, 139
Tanks, nonpetroleum, 175
Temperature measurement, 151
    gauging system, 17
Tightness testing methods, 139, 151, 197
Toluene, 175
Tracer methods, 30
Tracers, 131
Tubes, sensor, 123

U

U.S. Postal Service, 189

V

Valve pit leak, underground piping, 131
Vapor diffusion, 73
Vapor monitoring, 115, 123, 139
Vapor sensor testing, 90
Volatile tracers, 131
Volumetric leak detection, 3

X

Xylene, 90, 175